Data-Driven Design of Fault Diagnosis Systems

Adel Haghani Abandan Sari

Data-Driven Design of Fault Diagnosis Systems

Nonlinear Multimode Processes

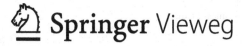

Adel Haghani Abandan Sari
Rostock, Germany

PhD Thesis, University of Duisburg-Essen, 2013

ISBN 978-3-658-05806-7   ISBN 978-3-658-05807-4 (eBook)
DOI 10.1007/978-3-658-05807-4

The Deutsche Nationalbibliothek lists this publication in the Deutsche Nationalbibliografie; detailed bibliographic data are available in the Internet at http://dnb.d-nb.de.

Library of Congress Control Number: 2014937515

Springer Vieweg
© Springer Fachmedien Wiesbaden 2014

Printed on acid-free paper

Springer Vieweg is a brand of Springer DE.
Springer DE is part of Springer Science+Business Media.
www.springer-vieweg.de

Preface

In many industrial applications early detection and diagnosis of abnormal behavior of the plant is of great importance. For decades, model-based methods have been widely used to design fault diagnosis systems. These approaches involve rigorous development of a process model based on first principles. During the last decades, the complexity of process plants has been drastically increased, which imposes great challenges in development of model-based monitoring approaches and it sometimes becomes unrealistic for modern large-scale processes. Alternative to model-based approaches, data-driven methods have been developed, which offer powerful tools to extract useful information for design of monitoring systems based on the available process measurements. Multivariate statistical process monitoring approaches are successfully applied for fault detection and diagnosis in technical processes.

However, many industrial processes are intrinsically nonlinear and operate in different operation regimes due to different product specifications, working environments and economic considerations. Due to the nonlinearity of processes, the performance of the classical multivariate statistical process monitoring methods, which are mainly based on the linearity assumption, becomes unsatisfactory, since the process characteristics will change from one operating point to another.

The main objective of the work presented here is to study and develop efficient fault diagnosis techniques for complex industrial systems, using process historical data and considering the nonlinear behavior of the process. The nonlinear system is assumed to be linear around the operating points and therefore considered as a piecewise linear system corresponding to each operating mode. To this end, different methods are presented to solve the fault detection problem based on the overall behavior of the process and its dynamics. Moreover, a novel technique is proposed for fault isolation and determination of the root cause of the faults in the system, based on the influence of the fault on the process measurements.

After successful detection and isolation of the fault, the faulty component in the system should be repaired or replaced. Moreover, there should be the possibility to temporarily remedy the fault consequences by changing the setting of a specific component in the system. Therefore, it is necessary to have a decision support system which can help the plant engineers to determine the proper maintenance operation after detection of a faulty component in the system. To address this problem, a methodology is proposed here which makes use of the results achieved in previous steps. Economic assessment of the possible maintenance operations is integrated into this system to provide the best operation in terms of the highest impact on process performance and the lowest losses.

The performance and effectiveness of the approaches proposed in this work are studied through their application on industrial benchmarks. A laboratory set up of a continuous stirred tank heater and the dryer section of a paper making machine are considered to carry out this study.

I would like to express deep gratitude to Prof. Dr.-Ing. Steven X. Ding for his consistent encouragement, guidance and support throughout my study at University of Duisburg-Essen. He was always a great source of inspiration, advice, help and support. My sincere appreciation must also go to Prof. Sirkka-Liisa Jämsä-Jounela, Department of Biotechnology and Chemical Technology at Aalto University, for her constructive comments on the manuscript of this work.

I would also like to thank all my friends and colleagues at institute of automatic control and complex systems (AKS), particularly Dr.-Ing. Birgit Köppen-Seliger and Dipl.-Ing. Eberhard Goldschmidt for their guidance and fruitful discussions in various situations and projects. My sincere thanks to Dipl.-Ing Jonas Esch for his great support during this work and I wish him the best for his future endeavors. I give my best wishes to M.Sc. Minjia Krüger, M.Sc. Haiyang Hao, Dipl.-Ing.

Christoph Kandler, M.Sc. Hao Luo,M.Sc Chris Louen and M.Sc. Tim Könings for thier research work and acknowledge their help in several occasions. A particular note of thanks is due to Mrs. Sabine Bay for her help in organizational responsibilities and Dipl.-Ing. Klaus Göbel for his technical support.

I am especially indebted to the staff at the institute of automation in university of Rostock, in particular Prof. Dr.-Ing Torsten Jeinsch for his support and patience.

Finally, I offer my heartfelt gratitude to my wife for her love and support over the years. I am most sincerely grateful to my family for their infinite kindness and incredible support throughout my life, without which I would never have come so far.

Adel Haghani

Contents

List of Notations

ARMAX	AutoRegressive Moving Average with eXogenous inputs
ARX	AutoRegressive eXogenous
BIC	Bayesian information criterion
CSTH	Continuous stirred tank heater
CVA	Canonical variate analysis
DO	Diagnostic observer
DPCA	Dynamic principal component analysis
DPLS	Dynamic partial least squares
EM	Expectation-Maximization
FD	Fault detection
FDD	Fault detection and diagnosis
FDF	Fault detection filter
FDI	Fault detection and isolation
FDIA	Fault detection, isolation and analysis
FGMM	Finite Gaussian mixture model
FTC	Fault tolerant control
GMM	Gaussian mixture model

ICA   Independent component analysis

KPCA   Kernel principal component analysis

LPV   Linear parameter varying

LTI   Linear time-invariant

MAP   Maximum *a posteriori* probability

MLE   Maximum likelihood estimate

MOESP   Multivariable output error state space

MPC   Model predictive controller

MSPM   Multivariate statistical process monitoring

N4SID   Numerical subspace state space system identification

NIPALS   Nonlinear iterative partial least squares algorithm

PCA   Principal component analysis

PDF   Probability density function

PI   Proportional integral

PLS   Partial least squares

PS   Parity space

PWA   Piecewise affine

PWARX   Piecewise autoregressive exogenous

RBC   Reconstruction based contribution

SIM   Subspace identification methods

SPE   Squared prediction error

SPM   Statistical process monitoring

SVC   Support vector classifier

SVD   Singular value decomposition

Mathematical symbols

$\mathbf{x}^T$	Transpose of $\mathbf{x}$
$\mathbf{X}^\dagger$	Pseudoinverse of $\mathbf{X}$
$\|\cdot\|$	2-norm
$\hat{x}$	Estimate of x
$\mathbb{R}$	Set of real numbers
$\mathbb{R}^n$	Set of n-dimensional real vectors
$\mathbb{R}^{n \times m}$	Set of $n \times m$ dimensional real matrices
$\in$	Belongs to
$p(x)$	Probability of state x
$p(x\|y)$	Conditional probability of x, given y
$I_{n \times n}$	$n \times n$ dimensional identity matrix

Control-theoretical symbols

$\mathbf{u}$	Input vector
$\mathbf{y}$	Output vector
$\mathbf{x}$	State vector
$\mathbf{v}$	Sensor noise
$\mathbf{w}$	Process disturbance
$\mathbf{d}$	Unknown disturbance
$\mathbf{f}$	Fault vector
l	Number of inputs
m	Number of outputs
n	Model order
$\mathbf{A}$	System matrix
$\mathbf{B}$	Input matrix
$\mathbf{C}$	Output matrix
$\mathbf{D}$	Feedthrough matrix

$\mathbf{E}_d$	Disturbance matrix
$\mathbf{F}_d$	Disturbance feedthrough
$\mathbf{E}_f$	Process fault distribution matrix
$\mathbf{F}_f$	Sensor fault distribution matrix
$\mathbf{L}$	Observer gain matrix
r	Residual signal
$\mathbf{v}_s$	Parity vector
$\mathbf{H}_{u,s}$	Input distribution matrix
s	Order of parity vector
J_{th}^I	Threshold with respect to index I

Statistical symbols

$\mathbf{\Sigma}$	Covariance matrix
$\boldsymbol{\mu}$	Mean vector
α	Confidence level
$\mathcal{N}$	Normal distribution
$\mathcal{U}$	Uniform distribution
F	F-distribution
χ^2	Chi-squared distribution
$E\{x\}$	Statistical expectation of x

List of Figures

List of Tables

1 Introduction

Technological developments in recent decades make the modern industrial and manufacturing systems more complex and pose increasing challenges in their design, analysis and integration. In such systems, asset management and equipment maintenance play a crucial role in enhancing the economic operation, product quality, overall system reliability and safety [58]. In a large scale system, such as the process industry, most of the losses and repairs are due to equipment and control system malfunction caused by aging, unanticipated interactions of components, misuse of components, etc [109]. Considering the scale and complexity of such systems, it becomes difficult for the plant engineer to detect the abnormality and find the causes of the abnormal events. In these situations, the human operator error is inevitable which may lead to more losses. The financial consequences of the system failures can be reduced, if they are detected and controlled in advance.

In a survey in U.S.A, it has been estimated that the petro-chemical industry loses 20 billion dollars per year due to process abnormalities [108]. Another instance which shows the significance of early detection of abnormality in industrial system is the incident happened recently (Sep. 24th, 2012) in Krefeld, Germany. The fire in a fertilizer manufacturing company causes a giant cloud of smokes that could be seen as far as Düsseldorf, Duisburg and other cities in the lower Rhine. Initial results of investigation show that a technical defect in material handling and conveyor systems was the cause of the incident [101]. Primary estimate shows that the financial losses due to the incident was tens of million euros, approximately [102].

In addition to financial aspects, there are many other incentives to develop methods for automatic detection, diagnosis and prognosis of malfunctions in the plant. That motivated academia to study and develop methods for detection and analysis of malfunctions in systems, known as **Fault diagnosis** in automatic control community, starting from early 70's [37].

1.1 The fault diagnosis problem

Before starting the discussion about the overall concept of fault diagnosis, let us describe what a fault means. According to the suggestion of IFAC[1] Technical Committee SAFEPROCESS [56]:

> "A fault is an unpermitted deviation of at least one characteristic property or parameter of the system from the acceptable/usual/standard condition."

The task of fault diagnosis consists of the following essential parts [38]:

- *Fault detection*: detection of the time of occurrence of faults in the functional units of the process, which lead to undesired or unacceptable behaviour of the whole system

- *Fault isolation*: localization of different faults

- *Fault analysis or identification*: determination of the type, magnitude and cause of the fault

A fault diagnosis system, depending on its functionality, is often called FD (for fault detection), FDI (for fault detection and isolation) or FDIA (for fault detection, isolation and analysis).

1.2 Classification of fault diagnosis methods

A traditional method for fault detection and diagnosis (FDD) is to use hardware redundancy by reconstructing the process component using

---

[1]International Federation of Automatic Control

the identical hardware component [38]. In this scheme a fault can be detected when there is an inconsistency between the outputs of process and the redundant component. Although this method provides direct isolation of the fault with high reliability, its application is limited to such safety-critical system as spacecrafts and nuclear power plants, due to extra cost and space required for the redundant component [108].

During the last 3 decades, advances in modern control theory led to powerful techniques for mathematical modeling, system identification and state estimation which have provided novel schemes for FDD, together with progresses of computer technology. The basic idea of these schemes is to replace the hardware redundancy by a process model in the form of a computer software which is known as analytical redundancy. The process model is a quantitative description of the process, obtained by first principles or its qualitative description. In order to detect faults in the system, same as hardware redundancy scheme, the plant outputs are compared with their estimation provided by the analytical model. The difference between the measured output and its estimation is called residual signal which is zero in fault-free case and deviates from zero in faulty case. The process of creating this signal is called residual generation. The residual signal delivers information about the faults, uncertainties and unknown disturbances which exist in the plant. To extract the useful information about the faults, post-processing of the residual signal is often carried out, which is called residual evaluation. Since these approaches are based on the model derived from the process, they are also known as model-based fault diagnosis techniques. The fundamental concept of the model-based fault diagnosis is shown in Fig. 1.1.

Most of the schemes developed for quantitative model-based FDD are based on the filtering and estimation techniques in modern control theory (e.g. Kalman filtering and observer-based techniques [5, 37, 88]). FDD methods based on on-line parameter estimation using least square method have been proposed to detect the drift in unknown parameters [55]. Another class of model-based FDD methods which has been extensively studied is parity space method whose core is

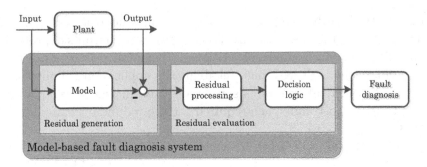

Figure 1.1: Model-based fault diagnosis scheme [29]

the representation of the system in the so called parity relation form [24, 80, 119].

Different from quantitative model-based scheme, in qualitative model-based methods the process model is derived from some fundamental physical knowledge of the system and represented in the form of causal relationship and if-then-else rules. The proper decision can be inferred using this qualitative model and given observations [109]. This class of FDD methods is very useful in large-scale system, where obtaining the quantitative process model is difficult or infeasible.

Nevertheless, in many large scale industrial processes, due to complexity of the system and large number of observations, deriving the quantitative and qualitative models based on first principles is not reasonable. In such cases, it is desired to model the process directly from the process input and output data. In industrial applications, the historical data which are collected routinely by process computers, contain hundred to thousand of variables measured every few seconds that are highly correlated to each other. Moreover, the statistical rank of data is very low and depends only on a few number of independent sources responsible for the variations in the process. It can be interpreted that the process is driven by small set of independent latent variables in the system. For that, several methods based on multivariate statistics [92, 67, 48, 50] have been developed, known as multivariate statistical process monitoring (MSPM) methods. The

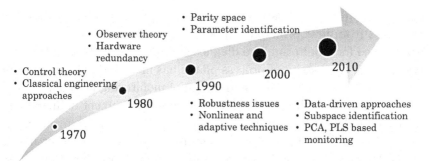

Figure 1.2: Historical development of FDD techniques

main objective of MSPM methods is to derive few meaningful measures from large amount of process measurements which reflect the
faulty behavior of the plant.

Generally, modern technical processes are nonlinear dynamic processes which are working under different operating regimes due to
different product specifications and constraints. The classical MSPM
methods cannot tackle this aspect of processes due to linearity assumption. Therefore combination of the model-based techniques and
statistical tools has gained more attention and merged with the filtering, estimation and identification theories. An intuitive way is to
estimate the system model using identification techniques and design
the FDI scheme using model-based techniques [96].

The first concern in design of FDI systems is detection performance,
i.e., the sensitivity of the residual signal to the faults and its robustness
against disturbances. To this end, in the last two decades numerous
efforts have been made to establish the effective FDI schemes and improve their sensitivity and robustness. The chronological development
of FDD techniques is depicted in Fig. 1.2.

1.3 Motivation and objective

In many industrial manufacturing and production applications, early
detection of faults and malfunctions which affect the product quality,

is of great interest. Based on analytical process models which are obtained from first principles, model based schemes have been well established for monitoring and diagnosis purposes [29]. During the last decades the complexity of process plants has increased, which imposes great challenges on modeling and design of monitoring methods. Since the model-based techniques involve rigorous development of analytical process models, their application on complex large scale plants is sometimes not feasible and becomes unrealistic. On the other hand, in modern technical processes huge amount of process variables are measured and recorded in process historian which can be used to design the monitoring systems. Using the historical process data, the well-established data-driven methods offer a powerful tool to extract useful information and discover the underlying structure of process.

Many industrial processes however are intrinsically nonlinear and operate in different operation regimes according to different product specifications, working environments and economic considerations. Due to the nonlinearity of processes, the performance of the classical multivariate statistical process monitoring methods, which are mainly based on the linearity assumption, becomes unsatisfactory, since the process model will change from one operating point to another. It leads to false alarms or missed detection of faults.

Based on the above mentioned observations, the main objective of the work presented here is to develop efficient fault diagnosis techniques for modern industrial systems, considering the plant's nonlinear behavior and its complexity. For that, the FDI problem is considered from plant overall performance perspective, i.e. the focus is on detection of those faults which are affecting the plant performance and production quality. Besides that, the following important practical issues are considered:

- Modeling of a complex industrial system requires excessive engineering efforts. The FDD techniques that are proposed in this work, are based on the available process historical data, which minimize the engineering efforts in the design step as well as uncertainties in the modeling.

- Most FDD techniques are based on the assumption that the system is linear time-invariant and therefore their applications are restricted to such systems. Although the plant's linearized model can be used for FDD design, in many applications the performance of these techniques will not be convincing due to set-point changes.

- FDD methods are designed to detect and diagnose the malfunctions in the process. From the enterprise's perspective, those faults and malfunctions which are affecting the plant performance, product quality and have financial consequences are of great importance. Thus the FDD system should be able to evaluate the effects of the faults on the plant performance and product quality and assess their financial impacts.

- It is often necessary to integrate the human expert knowledge into the diagnosis system. In this work, Bayesian inference technique has been used to incorporate process expert knowledge about the certain events in the system.

- A large scale industrial system is often equipped with many sensors and several thousands of signals are measured and recorded which imposes severe limitation on memory storage and on-line computation cost of the monitoring algorithm. Consequently, the monitoring algorithm should be designed efficiently with low computation and memory loads.

After successful detection and isolation of the fault, the faulty component in the system should be repaired or replaced. Moreover, there should be the possibility to temporarily remedy the fault consequences by changing the setting of a specific component in the system, e.g. retuning the controller and postponing the repair to the next plant shutdown to reduce the maintenance costs. Therefore, it is necessary to have a decision support system which can help the plant engineers to determine the faulty component and aid them to perform the proper corrective operation based on the current plant situation and constraints. To address this issue, a methodology is proposed here which integrates the FDD results with economical assessment of the corrective operations and their impact on the overall performance of the system.

1.4 Outlines

In this last section of this introductory chapter, the structure and contributions of this thesis will be briefly introduced. In Chapter 2, a brief overview of the representation of a technical system is given. Most common model-based FDD method, data-driven approaches and their combinations are discussed in this chapter. Fault detection filter, diagnostic observer, parity space-based approach and their relations are also discussed therein. Principal component analysis and partial least squares which are the popular approaches in statistical process monitoring, are also introduced in this chapter. Their strength and shortcoming from practical point of view are discussed and their recent extension and developments are introduced.

Chapter 3 deals with the performance-oriented FD problem in nonlinear stationary processes. For that, the nonlinear system is considered as piecewise linear system according to each operating mode and a novel method for detection of the faults, which are affecting the product quality in such multimode systems, is proposed. A new FD index is employed which represents the probability that a fault is happening in the system with focus on those faults which affect the product quality.

Chapter 4 describes the further application of the methodology developed in Chapter 3 for the cases where the process under consideration is a dynamic system. The problem of dynamic changes in multimode systems is addressed and a new approach for fault detection in nonlinear multimode dynamic systems is proposed. The parity space representation of the system is identified using the historical data and based on it a multi observer-based residual generation scheme is developed. A Bayesian inference technique is further utilized to combine the local results and develop a global index for fault detection purpose.

The focus of Chapter 5 is on fault isolation. After detection of a fault in the system, it is crucial to isolate the fault before performing the corrective operations. An overview of the classical fault isolation namely contribution analysis and reconstruction-based contribution analysis is given at the beginning. Furthermore a new probabilistic approach is proposed to address the fault isolation problem in nonlinear multimode systems.

Chapter 6 discusses the last element in process monitoring chain which is a decision support system. After successful fault detection and isolation, the system has to be recovered from the abnormal situation. The faulty component should be repaired or replaced. The task of decision support system is to guide the human operator to perform the most proper corrective operation based on the assessment of the current plant situation and constraints. To this aim, in this chapter a decision support system is proposed which integrates the FDI results with economical evaluation of possible corrective operation to find the best possible correction with high impact on plant overall performance and low costs.

In Chapter 7, the algorithm developed in previous chapters are demonstrated through benchmark examples. The proposed approaches are partially compared with existing methods and their performance and effectiveness are discussed. For this purpose, a laboratory setup of continuous stirred tank heater is considered. To evaluate the results in industrial scale, the methods are also demonstrated using the data obtained from a paper production machine.

Chapter 8 summarizes the works which have been done and gives an overview on the future directions. For the sake of clarity, the organization of the chapters is shown in Fig. 1.3.

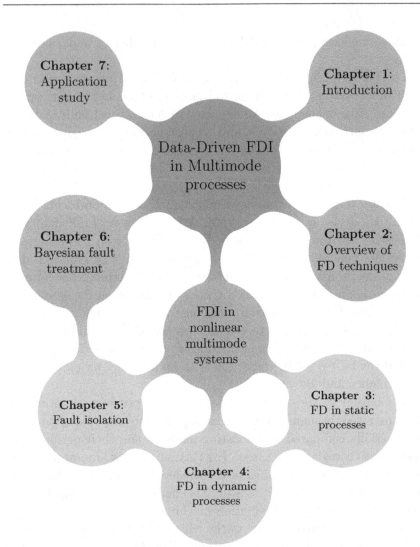

Figure 1.3: Organization of the chapters

2 An overview of fault diagnosis techniques

Generally speaking, the main objective of FDD is to convert the process observations into a few measures, indicating the anomalies of the process behavior and thereby help the plant engineer in determining the status of the plant. An intuitive approach might be limit sensing, i.e. defining thresholds for signals and raising alarms when signals cross the thresholds [96]. Despite its simplicity, limit sensing has several drawbacks, e.g. it does not consider the interaction between different components of the plant and the correlation between the observations. Moreover, setting thresholds and monitoring the measurements in a large scale process which e.g. consists of several thousands of variables is not feasible.

Another method for providing information indicating anomalies in the plant behavior is to use redundancy in hardware or analytical form. The core of the methods based on analytical redundancy is a process model running parallel to the process, which provides estimates of the process outputs. These models were derived through rigorous approaches, e.g. first principles. However, obtaining mathematical process model for large scale complex industrial process is a challenging task.

Thanks to the recent developments, modern industrial processes are becoming more and more instrumented. Large amount of data are measured and recorded in process historian which can be utilized for data-driven design of FDD system.

The first part of this chapter gives an overview of the available model-based FDD methods and their features. Later, the data-driven methods for process monitoring and FDD are discussed. But before starting the discussion about model-based FDD methods, let us give a short overview about the representation of the process model used throughout this thesis.

2.1 Representation of the technical process

Linear time-invariant (LTI) systems are the most important representation of dynamical systems considered in practice. The time domain realization of the LTI systems can be expressed in different forms, e.g. AutoRegressive eXogenous (ARX), AutoRegressive Moving Average with eXogenous inputs (ARMAX) [78]. For instance, the ARX representation of an LTI system is given below.

$$y(k) = \sum_{i=1}^{n_a} a_i y(k-i) + \sum_{j=1}^{n_b} b_i u(k-i). \tag{2.1}$$

Nevertheless, the most useful representation of dynamic systems in time domain is state space representation, since the physical knowledge of system can be more effectively incorporated into the state space model. In state space model the relation between process inputs, noise and outputs is described by means of a system of first order ordinary differential equations via a state vector. The standard form of state space description of a discrete-time LTI system is given by

$$\begin{aligned} \mathbf{x}(k+1) &= \mathbf{A}\mathbf{x}(k) + \mathbf{B}\mathbf{u}(k) \\ \mathbf{y}(k) &= \mathbf{C}\mathbf{x}(k) + \mathbf{D}\mathbf{u}(k), \end{aligned} \tag{2.2}$$

where $\mathbf{x}(k) \in \mathbb{R}^n$ is the state vector at discrete time instant k with initial condition $\mathbf{x}(0) = \mathbf{x}_0$, $\mathbf{u}(k) \in \mathbb{R}^l$ the input vector and $\mathbf{y}(k) \in \mathbb{R}^m$ the output vector at instant k. The matrix $\mathbf{A} \in \mathbb{R}^{n \times n}$ is called system matrix which describes the eigen dynamics of the system. $\mathbf{B} \in \mathbb{R}^{n \times l}$ is input matrix which represents the linear transformation by which the input vector affects the state vector at instant $k+1$. $\mathbf{C} \in \mathbb{R}^{m \times n}$ shows how the internal states transferred to the process output and the term with $\mathbf{D} \in \mathbb{R}^{m \times l}$ is called direct feedthrough term [107]. The

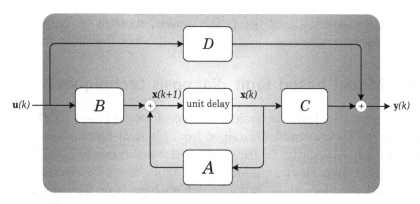

Figure 2.1: Block diagram of state space representation

block diagram of state space representation in Eq. (2.2) is plotted in Fig. 2.1.

In practice, environmental disturbances often cause unexpected changes in the process and the process measurements are inevitably contaminated by noises. The process disturbances and noises can be integrated into the state space equation in Eq. (2.2) as follows:

$$\mathbf{x}(k+1) = \mathbf{A}\mathbf{x}(k) + \mathbf{B}\mathbf{u}(k) + \mathbf{E}_d\mathbf{d}(k) + \mathbf{w}(k)$$
$$\mathbf{y}(k) = \mathbf{C}\mathbf{x}(k) + \mathbf{D}\mathbf{u}(k) + \mathbf{F}_d\mathbf{d}(k) + \mathbf{v}(k), \qquad (2.3)$$

where $\mathbf{d}(k) \in \mathbb{R}^{k_d}$ is the unknown disturbance vector and $\mathbf{E_d}$ and $\mathbf{F_d}$ are constant matrices with appropriate dimensions. Vectors $\mathbf{w}(k)$ and $\mathbf{v}(k)$ are unmeasurable white noise sequences.

It is also interesting to introduce the fault in the modeling of the process. Integration of the fault in the state space model in Eq. (2.2) is given by

$$\mathbf{x}(k+1) = \mathbf{A}\mathbf{x}(k) + \mathbf{B}\mathbf{u}(k) + \mathbf{E}_f\mathbf{f}(k)$$
$$\mathbf{y}(k) = \mathbf{C}\mathbf{x}(k) + \mathbf{D}\mathbf{u}(k) + \mathbf{F}_f\mathbf{f}(k). \qquad (2.4)$$

In Eq. (2.4), $\mathbf{f} \in \mathbb{R}^{k_f}$ is the fault vector and $\mathbf{E}_f$ and $\mathbf{F}_f$ are fault distribution matrices with appropriate dimensions. $\mathbf{E}_f$ and $\mathbf{F}_f$ represent the location where a fault occurs (i.e. sensor, actuator or the process) and the way it affects the system (i.e. additive or multiplicative).

2.2 Model-based fault diagnosis techniques

As discussed earlier, the basic principle of model-based FDD techniques is a process model obtained from physical description of the system. In the early 1970s, motivated by the newly established observer theory, the first model-based fault detection method was proposed [6]. Since then, various methods have been developed and reported, for instance diagnostic observer, parity space methods and their performance and robustness has been studied. From the industrial perspective, they have also gained tremendous attention and been used for different applications. In order to keep it short, among the existing model-based FDD techniques, fault detection filter, diagnostic observer-based and parity space-based residual generators which have received more attention from academic and industrial point of view, are introduced here. Moreover, their interconnection, comparison and some remarks are included.

2.2.1 Fault detection filter

The observer-based residual generation techniques originated from the pioneering works of Beard [6] and Jones [63], known as fault detection filter (FDF). The construction of FDF has been achieved by the design of a full-order state observer. Consider a discrete LTI system described by state space equation in Eq. (2.2). The FDF for the system can be constructed as

$$\hat{\mathbf{x}}(k+1) = \mathbf{A}\hat{\mathbf{x}}(k) + \mathbf{B}\mathbf{u}(k) + \mathbf{L}(\mathbf{y}(k) - \hat{\mathbf{y}}(k))$$
$$\hat{\mathbf{y}}(k) = \mathbf{C}\hat{\mathbf{x}}(k) + \mathbf{D}\mathbf{u}(k), \tag{2.5}$$

where $\mathbf{L}$ is the observer gain and chosen such that $\mathbf{A} - \mathbf{L}\mathbf{C}$ is stable (i.e. its eigenvalues are located inside the unit circle). In this case

$$\lim_{k \to \infty} (\mathbf{x}(k) - \hat{\mathbf{x}}(k)) = 0. \tag{2.6}$$

It is worth mentioning that the proper selection of $\mathbf{L}$ can strongly affect the performance of estimation. By choosing $\mathbf{e}(k) = \mathbf{x}(k) - \hat{\mathbf{x}}(k)$, the dynamics of estimation error in residual generator is described by

$$\mathbf{e}(k+1) = (\mathbf{A} - \mathbf{LC})\mathbf{e}(k)$$
$$\mathbf{r}(k) = \mathbf{Ce}(k), \tag{2.7}$$

where $\mathbf{r}$ is the residual signal and defined as $\mathbf{r}(k) = \mathbf{y}(k) - \hat{\mathbf{y}}(k)$. In practice, it is often useful to use a post-filter to improve the sensitivity and robustness of FDF as follows

$$\mathbf{r}(k) = \mathbf{V}(\mathbf{y}(k) - \hat{\mathbf{y}}(k)). \tag{2.8}$$

As the result, the design of FDF can be summarized as optimal selection of observer gain $\mathbf{L}$ and post-filter $\mathbf{V}$ to achieve high estimation performance, sensitivity to faults and robustness against disturbances. The main drawback of FDF is due to on-line implementation of a full-order state observer, since in many applications a reduced order observer may provide the same information for FDD purpose.

2.2.2 Diagnostic observer

Basically, diagnostic observer (DO) is a form of Luenberger (output) observer which is used for residual generation purpose. The Luenberger observer is described by [79]

$$\mathbf{z}(k+1) = \mathbf{A}_z\mathbf{z}(k) + \mathbf{B}_z\mathbf{u}(k) + \mathbf{L}_z\mathbf{y}(k)$$
$$\hat{\mathbf{y}}(k) = \bar{\mathbf{C}}_z\mathbf{z}(k) + \bar{\mathbf{D}}_z\mathbf{u}(k) + \bar{\mathbf{G}}_z\mathbf{y}(k) \tag{2.9}$$

where $\mathbf{z} \in \mathbb{R}^s$ with s representing the order of observer that is equal to or less than the order of system n. The design of observer in Eq. (2.9) is achieved by solving the so-called Luenberger equations:

- $\mathbf{A}_z$ is stable

- $\mathbf{TA} - \mathbf{A}_z\mathbf{T} = \mathbf{L}_z\mathbf{C}$, $\mathbf{B}_z = \mathbf{TB} - \mathbf{L}_z\mathbf{D}$

- $\mathbf{C} = \bar{\mathbf{C}}_z\mathbf{T} + \bar{\mathbf{G}}_z\mathbf{C}$, $\bar{\mathbf{D}}_z = -\bar{\mathbf{G}}_z\mathbf{D} + \mathbf{D}$, where $\mathbf{T}$ is the state transformation matrix. Considering $\mathbf{e}(k) = \mathbf{Tx}(k) - \mathbf{z}(k)$ as the observer estimation error, its dynamics is governed by

$$\mathbf{e}(k+1) = \mathbf{A}_z\mathbf{e}(k)$$
$$\mathbf{y}(k) - \hat{\mathbf{y}}(k) = \bar{\mathbf{C}}_z\mathbf{e}(k), \qquad (2.10)$$

which provides an unbiased estimate of the output signal.

For FDD purpose, the residual generator can be constructed in the following form:

$$\mathbf{z}(k+1) = \mathbf{A}_z\mathbf{z}(k) + \mathbf{B}_z\mathbf{u}(k) + \mathbf{L}_z\mathbf{y}(k)$$
$$r(k) = \mathbf{g}_z\mathbf{y}(k) - \mathbf{c}_z\mathbf{z}(k) - \mathbf{d}_z\mathbf{u}(k), \qquad (2.11)$$

where $r \in \mathbb{R}$ serves as the residual signal and $\mathbf{g}_z = V(\mathbf{I} - \bar{\mathbf{G}}_z)$, $\mathbf{c}_z = V\bar{\mathbf{C}}_z$, $\mathbf{d}_z = V\bar{\mathbf{D}}_z$. In this case the 3rd Luenberger condition should be replaced with

$$VC - \mathbf{g}_z\mathbf{T} = \mathbf{0}, \ \mathbf{d}_z = VD. \qquad (2.12)$$

Compared to FDF, the main advantage of DO is simple on-line implementation through a reduced order observer and lower computation cost. During the past 30 years, large number of algorithms have been developed for solving the Luenberger equations. Moreover, the existence conditions, minimum order of observer and parametrization of the solution have been intensively studied [29].

2.2.3 Parity space approach

In this section, parity space (PS) approach for residual generation is described. In this approach the so-called parity relation is used instead of the observer for residual generation purpose. Consider the state space model of a system shown in Eq. (2.2). Using the past s input and output measurements, the state space equations can be described in following form [80]

$$\mathbf{y}_s(k) = \mathbf{\Gamma}_s\mathbf{x}(k - s + 1) + \mathbf{H}_{u,s}\mathbf{u}_s(k) + \mathbf{H}_{d,s}\mathbf{d}_s(k), \qquad (2.13)$$

known as parity relation where

$$\mathbf{y}_s(k) = \begin{bmatrix} \mathbf{y}(k-s+1) \\ \mathbf{y}(k-s+2) \\ \vdots \\ \mathbf{y}(k) \end{bmatrix} \in \mathbb{R}^{sm}, \ \mathbf{u}_s(k) = \begin{bmatrix} \mathbf{u}(k-s+1) \\ \mathbf{u}(k-s+2) \\ \vdots \\ \mathbf{u}(k) \end{bmatrix} \in \mathbb{R}^{sl}$$

$$\mathbf{\Gamma}_s = \begin{bmatrix} \mathbf{C} \\ \mathbf{CA} \\ \vdots \\ \mathbf{CA}^{s-1} \end{bmatrix} \in \mathbb{R}^{sm \times n}, \quad \mathbf{d}_s(k) = \begin{bmatrix} \mathbf{d}(k-s+1) \\ \mathbf{d}(k-s+2) \\ \vdots \\ \mathbf{d}(k) \end{bmatrix} \in \mathbb{R}^{sk_d},$$

$$\mathbf{H}_{u,s} = \begin{bmatrix} \mathbf{D} & \mathbf{0} & \cdots & \mathbf{0} \\ \mathbf{CB} & \mathbf{D} & \ddots & \vdots \\ \vdots & \ddots & \ddots & \mathbf{0} \\ \mathbf{CA}^{s-2}\mathbf{B} & \cdots & \mathbf{CB} & \mathbf{D} \end{bmatrix} \in \mathbb{R}^{sm \times sl}$$

$$\mathbf{H}_{d,s} = \begin{bmatrix} \mathbf{F}_d & \mathbf{0} & \cdots & \mathbf{0} \\ \mathbf{CE_d} & \mathbf{F}_d & \ddots & \vdots \\ \vdots & \ddots & \ddots & \mathbf{0} \\ \mathbf{CA}^{s-2}\mathbf{E}_d & \cdots & \mathbf{CE_d} & \mathbf{F}_d \end{bmatrix} \in \mathbb{R}^{sm \times sk_d}. \tag{2.14}$$

with s denoting the order of parity space. The parity relation in Eq. (2.13) describes the relationship between the process inputs and outputs incorporating the past state of the system.

To remove the term related to the past state vector $\mathbf{x}(k-s+1)$ in Eq. (2.13), consider $s \geq n$ and assuming that the following rank condition holds

$$rank(\mathbf{\Gamma}_s) = n \tag{2.15}$$

and the pair $(\mathbf{C}, \mathbf{A})$ is observable, there exists at least a row vector $\mathbf{v}_s \in \mathbb{R}^{sm}(\neq \mathbf{0})$ such that

$$\mathbf{v}_s \mathbf{\Gamma}_s = \mathbf{0}. \tag{2.16}$$

The vector $\mathbf{v}_s$ in Eq. (2.16) lies in the left null space of $\boldsymbol{\Gamma}_s$ and plays a central role in parity space approach and is often known as parity vector. The space spanned by parity vectors satisfying Eq. (2.16) is called parity space

$$\mathcal{P}_s = \{\mathbf{v}_s | \mathbf{v}_s \boldsymbol{\Gamma}_s = 0\}. \tag{2.17}$$

Neglecting the term representing the disturbances in Eq. (2.13), the parity space-based residual generator is constructed by

$$r(k) = \mathbf{v}_s(\mathbf{y}_s(k) - \mathbf{H}_{u,s}\mathbf{u}_s(k)). \tag{2.18}$$

The residual signal $r(k)$ in Eq. (2.18) is equal to zero in fault- and disturbance-free case. In general, the design form is

$$r(k) = \mathbf{v}_s(\mathbf{H}_{d,s}\mathbf{d}_s(k) + \mathbf{H}_{f,s}\mathbf{f}_s(k)), \tag{2.19}$$

where $\mathbf{H}_{f,s}$ and $\mathbf{f}_s(k)$ are constructed in similar form as represented in Eq. (2.14). Equation (2.19) shows that the residual signal depends on the fault and disturbances. The sensitivity and robustness of the parity space approach have been extensively studied. For more details, the readers are referred to [14, 29].

The construction of PS-based residual generator is straightforward compared to the observer-based one. The design step includes calculation of parity vectors via solving the linear optimization problem in Eq. (2.16). The on-line implementation of PS-based approach involves considering the past and temporal data, which is a shortcoming compared to the observer-based methods.

2.2.4 Relationship between PS and DO

Diagnostic observer and parity space-based method are two powerful methods often used for residual generation. Studies on the interconnections and comparison of different residual generation techniques revealed an interesting one-to-one relationship between design parameters of PS and DO methods [29, 137]. That means, the design parameters of the Luenberger observer in Eq. (2.11) can be obtained from parity vector and vice-versa. It has been shown in [31] that given the parity vector $\mathbf{v}_s = \begin{bmatrix} \mathbf{v}_{s,0} & \mathbf{v}_{s,1} & \cdots & \mathbf{v}_{s,s-1} \end{bmatrix}$, the parameters of DO can be obtained by

$$
\mathbf{A}_z = \begin{bmatrix} 0 & 0 & \cdots & 0 & 0 \\ 1 & 0 & \cdots & 0 & 0 \\ \vdots & \vdots & \ddots & \vdots & \vdots \\ 0 & 0 & \cdots & 1 & 0 \end{bmatrix}, \quad \mathbf{L}_z = - \begin{bmatrix} \mathbf{v}_{s,0} \\ \mathbf{v}_{s,1} \\ \vdots \\ \mathbf{v}_{s,s-2} \end{bmatrix}
$$

$$
\mathbf{c}_z = \begin{bmatrix} 0 & \cdots & 0 & 1 \end{bmatrix}, \quad \mathbf{g}_z = \mathbf{v}_{s,s-1}
$$

$$
\mathbf{d}_z = \mathbf{g}\mathbf{D}, \quad \mathbf{B}_z = \begin{bmatrix} \mathbf{v}_{s,0} & \mathbf{v}_{s,1} & \cdots & \mathbf{v}_{s,s-1} \\ \mathbf{v}_{s,1} & \mathbf{v}_{s,2} & \cdots & 0 \\ \vdots & \ddots & \ddots & \vdots \\ \mathbf{v}_{s,s-1} & 0 & \cdots & 0 \end{bmatrix} \begin{bmatrix} \mathbf{D} \\ \mathbf{CB} \\ \mathbf{CAB} \\ \cdots \\ \mathbf{CA}^{s-2}\mathbf{B} \end{bmatrix}
$$

$$
\mathbf{T} = \begin{bmatrix} \mathbf{v}_{s,1} & \mathbf{v}_{s,2} & \cdots & \mathbf{v}_{s,s-1} \\ \mathbf{v}_{s,2} & \mathbf{v}_{s,3} & \cdots & 0 \\ \vdots & \ddots & \ddots & \vdots \\ \mathbf{v}_{s,s-1} & 0 & \cdots & 0 \end{bmatrix} \begin{bmatrix} \mathbf{C} \\ \mathbf{CA} \\ \cdots \\ \mathbf{CA}^{s-2} \end{bmatrix}. \tag{2.20}
$$

Alternatively, given the matrices of a diagnostic observer $\mathbf{A}_z, \mathbf{L}_z, \mathbf{T}, \mathbf{c}_z$ and $\mathbf{g}_z$, the parity vector $\mathbf{v}_s$ can be obtained by

$$
\mathbf{v}_{s,s-1} = \mathbf{g}_z, \quad \begin{bmatrix} \mathbf{v}_{s,0} \\ \mathbf{v}_{s,1} \\ \vdots \\ \mathbf{v}_{s,s-2} \end{bmatrix} = -\mathbf{L}_z. \tag{2.21}
$$

This one-to-one relationship discloses the fact that the simple off-line design step in PS-based approach can be combined with the efficient on-line implementation of DO. In this scheme, the parity vector is obtained in off-line design step and then the parameters of diagnostic observer are directly calculated using Eq. (2.20). Compared to the PS-based residual generation, in this method the on-line implementation is carried out using the temporal data. Furthermore, the DO provides the possibility to arbitrarily select the poles of $\mathbf{A}_z$ to achieve the desired estimation performance.

2.3 Statistical process monitoring

Model-based FDD assumes the availability of the quantitative process model, which is not always the case in modern technical processes. Instead of that, the plant historical data can be used to build a statistical model for FDD purpose. In recent years, taking the advantages of progresses in technology and computer science, various schemes have been developed for monitoring of processes based on different statistical approaches known as statistical process monitoring (SPM) methods. The available SPM techniques are ranging from simple limit sensing to advanced time series analysis, classification and regression methods. Their applications have been expanded to different fields such as chemometrics and process control [96].

Methods based on limit sensing which determine thresholds for each observation ignore the serial and spatial correlation in measurements. To handle spatial correlation, monitoring methods based on principal component analysis (PCA) have been developed. PCA is a dimension reduction technique which considers the correlation among the process variables and captures the most variations in the data. Another basic SPM method is partial least squares (PLS) method. PLS method has been used in process monitoring in a similar way as PCA. PLS attempts to decompose the data in such a way that the correlation between predictor and predicted variables are maximized. In this section, these two common statistical process monitoring approaches and their applications and extensions are reviewed.

2.3.1 Principal component analysis

PCA is an optimal linear dimensionality reduction technique in terms of capturing the most variations in data. It reduces the dimension of monitoring space by projecting the data on a set of few orthogonal vectors known as loading vectors which explains the most variance in the data, without losing important information. Its ability to handle huge amount of highly correlated data and its simple realization make it a popular method for industrial process monitoring [94, 110].

The off-line design or training step of PCA based process monitoring is basically an eigenvalue-eigenvector problem. In the training step, the PCA model is obtained from the available fault-free measurements of the system. Let $\mathbf{X} \in \mathbb{R}^{N \times m}$ represent the N samples of m process measurements being used as training data. The data matrix $\mathbf{X}$ is scaled to zero mean and optionally to unit variance. Mean centering is an important step in statistical methods. Otherwise, the loading vectors do not describe the largest directions of variations in data. Rather than, it shows the combination of mean direction and direction of largest variation (see Fig. 2.2). The PCA algorithm decomposes $\mathbf{X}$ into two orthogonal subspaces

$$\mathbf{X} = \mathbf{T}_{pc}\mathbf{P}_{pc}^T + \mathbf{T}_{res}\mathbf{P}_{res}^T = \begin{bmatrix} \mathbf{T}_{pc} & \mathbf{T}_{res} \end{bmatrix} \begin{bmatrix} \mathbf{P}_{pc} & \mathbf{P}_{res} \end{bmatrix}^T = \mathbf{T}\mathbf{P}^T, \tag{2.22}$$

where the subspace spanned by $\mathbf{P}_{pc}$ is known as principal component subspace and the subspace spanned by $\mathbf{P}_{res}$ is known as residual subspace. From the fact that, the columns of $\mathbf{T}$ are orthogonal [92], $\mathbf{P}$ can be obtained by performing a singular value decomposition (SVD) on the covariance matrix as follows

$$\Sigma_{x,x} = \frac{1}{N-1}\mathbf{X}^T\mathbf{X} = \mathbf{P}\mathbf{\Lambda}\mathbf{P}^T, \tag{2.23}$$

where $\mathbf{\Lambda} = \mathrm{diag}(\sigma_1^2, \cdots, \sigma_m^2)$ is a diagonal matrix and contains the eigenvalues of covariance matrix in descending order. The $\mathbf{P}_{pc} \in \mathbb{R}^{m \times a}$ in Eq. (2.22) can be set as the columns of $\mathbf{P}$ corresponding to the a largest singular values of covariance matrix and $\mathbf{P}_{res} \in \mathbb{R}^{m \times (m-a)}$ corresponding to the $m - a$ smallest ones. The number of principal components a, can be determined using certain criteria e.g. cross validation test [121]. The decomposition in Eq. (2.22) for the case where $m = 2$ and $a = 1$ is shown in Fig. 2.2, for centered mean and non-centered mean data.

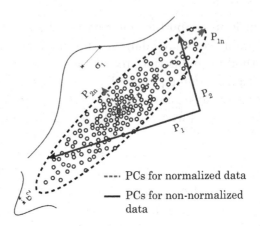

Figure 2.2: Principal component analysis

Every new normalized sample of measurements, $\mathbf{x}$, can be projected onto the principal and residual subspaces as

$$\hat{\mathbf{x}} = \mathbf{P}_{pc}\mathbf{P}_{pc}^{T}\mathbf{x}$$
$$\tilde{\mathbf{x}} = \mathbf{P}_{res}\mathbf{P}_{res}^{T}\mathbf{x} = (\mathbf{I} - \mathbf{P}_{pc}\mathbf{P}_{pc}^{T})\mathbf{x} \tag{2.24}$$

where $\mathbf{x} = \hat{\mathbf{x}} + \tilde{\mathbf{x}}$. For monitoring purpose, the so-called T^2 statistic and squared prediction error (SPE) can be employed

$$T^2 = \mathbf{x}^T\mathbf{P}_{pc}\mathbf{\Lambda}_{pc}^{-1}\mathbf{P}_{pc}^{T}\mathbf{x}$$
$$SPE = \mathbf{x}^T\mathbf{P}_{res}\mathbf{P}_{res}^{T}\mathbf{x}. \tag{2.25}$$

The fault will be detected if the indices exceed the corresponding thresholds:

$$J_{th}^{T^2} = \frac{a(N^2-1)}{N(N-a)}F_{\alpha}(a, N-a)$$
$$J_{th}^{SPE} = \theta_1\left(\frac{c_\alpha\sqrt{2\theta_2 h_0^2}}{\theta_1} + 1 + \frac{\theta_2 h_0(h_0-1)}{\theta_1^2}\right)^{1/h_0} \tag{2.26}$$

for T^2 and SPE indices respectively, where

$$\theta_i = \sum_{j=l+1}^{m} (\sigma_j^2)^i, \quad i = 1, 2, 3$$

$$h_0 = 1 - \frac{2\theta_1\theta_3}{3\theta_2^2} \tag{2.27}$$

and c_α is the normal deviate corresponding to the upper $1 - \alpha$ percentile. In order to simplify the threshold, assuming that the number of training samples N is large enough, the F-distribution in Eq. (2.26) can be approximated through a χ^2 distribution with a degrees of freedom and confidence level α, i.e.

$$J_{th}^{T^2} = \chi_\alpha^2(a). \tag{2.28}$$

To simplify the monitoring of the residual subspace, in [30] a new test statistic has been introduced which avoids the complexity of SPE index and its threshold:

$$T_{new}^2 = \mathbf{x}^T \mathbf{P}_{res}\Xi\mathbf{P}_{res}^T\mathbf{x}, \tag{2.29}$$

where

$$\Xi = \text{diag}\left(\frac{\sigma_m^2}{\sigma_{a+1}^2}, \cdots, \frac{\sigma_m^2}{\sigma_{m-1}^2}, 1\right). \tag{2.30}$$

For a given confidence level α the corresponding threshold can be calculated as

$$J_{th}^{T_{new}^2} = \sigma_m^2 \chi_\alpha^2(m - a). \tag{2.31}$$

The performance, sensitivity and fault detectability of this new index have been studied and compared with the classical indices in [127].

The application of the PCA method is restricted to the cases where the data follow uni-modal multivariate Gaussian distribution. However in many industrial applications the data are non-Gaussian due to nonlinearities in the process. For monitoring nonlinear processes, multivariate SPM methods based on kernel PCA (KPCA) and independent component analysis (ICA) have been developed. See their applications in [22, 68, 75, 138]. Moreover PCA captures the spatial correlation among measurements and ignores the serial correlation that happens often in processes which have dynamic behavior. The PCA method can be extended to include serial correlation by augmenting the temporal data at each instant to a sequence of past data. This approach is known as dynamic PCA (DPCA) [73]. For dynamic alternatives of KPCA and ICA, the readers are referred to [23, 51, 76].

2.3.2 Partial least squares

Partial least squares (PLS), also known as projection to latent structure is another multivariate statistical method adopted for process monitoring purpose. Compared to the PCA which captures the correlation among the process variables and finds the subspace which contains the most variations, PLS is used to determine the subspace which is correlated with the predicted block and describes the most variations in the predictor block. One application of PLS is to consider the process measurements $\mathbf{X}$ as the predictor block and the product quality measurements $\mathbf{Y}$ as the predicted block [96]. The product quality is often measured off-line in the laboratory and is not available on-line. PLS can be used for prediction of the product quality as well as process monitoring [66, 98, 120].

The steps involved in design and implementation of PLS method is similar to PCA method. The PLS algorithm involves projecting $\mathbf{X} \in \mathbb{R}^{N \times l}$ and $\mathbf{Y} \in \mathbb{R}^{N \times m}$ onto the latent variables $\mathbf{T} \in \mathbb{R}^{N \times \gamma}$ to build the correlation model of $\mathbf{X}$ and $\mathbf{Y}$,

$$\mathbf{X} = \mathbf{T}\mathbf{P}^{\mathbf{T}} + \tilde{\mathbf{X}} = \hat{\mathbf{X}} + \tilde{\mathbf{X}}$$
$$\mathbf{Y} = \mathbf{T}\mathbf{Q}^{\mathbf{T}} + \mathbf{E_y} = \mathbf{X}\mathbf{M} + \mathbf{E_y} \tag{2.32}$$

where γ is the number of latent variables and $\mathbf{P} \in \mathbb{R}^{l \times \gamma}$ and $\mathbf{Q} \in \mathbb{R}^{m \times \gamma}$ are loading matrices of $\mathbf{X}$ and $\mathbf{Y}$. The matrix $\mathbf{M} \in \mathbb{R}^{l \times m}$ is known as the regression coefficient. Typically the number of latent variables γ is

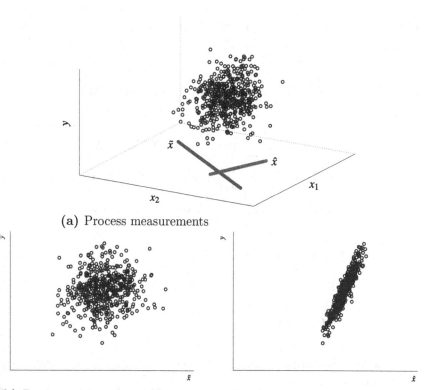

(a) Process measurements

b) Decomposition of variable spaces in PLS method

Figure 2.3: PLS method decomposes the predictor block **X** with respect to its correlation to predicted block **Y**.

determined by using cross-validation test [121]. In other words, PLS decomposes the subspace of regressor **X**, into two subspaces $\hat{\mathbf{X}}$ and $\tilde{\mathbf{X}}$, with respect to their correlation to dependant variables **Y**. The decomposition of **X** with respect to **Y** is shown in Fig. 2.3. Similar to PCA based process monitoring method , the T^2 and SPE indices are used for monitoring of these two subspaces.

The standard PLS approach for modeling the industrial process is formulated as follows, known as nonlinear iterative partial least squares algorithm (NIPALS) [25, 48, 50]:

- Collect N samples of $\mathbf{x}$ and $\mathbf{y}$ and normalize them to zero mean and unit variance to build $\mathbf{X} = \begin{bmatrix} \mathbf{x}_1 & \cdots & \mathbf{x}_l \end{bmatrix} \in \mathbb{R}^{N \times l}$ and $\mathbf{Y} = \begin{bmatrix} \mathbf{y}_1 & \cdots & \mathbf{y}_m \end{bmatrix} \in \mathbb{R}^{N \times m}$

- Perform γ times the following iterative computation:

  for $k = 1, \cdots, \gamma$:

$$(\mathbf{w}_k^*, \mathbf{q}_k^*) = \arg \max_{\|\mathbf{w}_k\|=1, \|\mathbf{q}_k\|=1} \mathbf{w}_k^T \mathbf{X}_k^T \mathbf{Y} \mathbf{q}_k, \quad \mathbf{X}_1 = \mathbf{X}$$

$$\mathbf{t}_k = \mathbf{X}_k \mathbf{w}_k^*, \quad \mathbf{p}_k = \frac{\mathbf{X}_k^T \mathbf{t}_k}{\|\mathbf{t}_k\|^2}, \quad \mathbf{X}_{k+1} = \mathbf{X}_k - \mathbf{t}_k \mathbf{p}_k^T$$

$$\mathbf{r}_1 = \mathbf{w}_1^*, \quad \mathbf{r}_k = \prod_{j=1}^{k-1} (\mathbf{I}_{p \times p} - \mathbf{w}_j^* \mathbf{p}_j^T) \mathbf{w}_k^*, \quad k > 1$$

- Calculate $\mathbf{P}$, $\mathbf{Q}$, $\mathbf{R}$, $\mathbf{T}$ and $\mathbf{M}$ using

$$\mathbf{P} = \begin{bmatrix} p_1 & \cdots & p_\gamma \end{bmatrix}, \quad \mathbf{T} = \begin{bmatrix} t_1 & \cdots & t_\gamma \end{bmatrix}$$
$$\mathbf{Q} = \begin{bmatrix} q_1 & \cdots & q_\gamma \end{bmatrix}, \quad \mathbf{R} = \begin{bmatrix} r_1 & \cdots & r_\gamma \end{bmatrix}$$
$$\mathbf{M} = \mathbf{R}\mathbf{Q}^T$$

To detect the faults in $\hat{\mathbf{X}}$ and $\tilde{\mathbf{X}}$, the T^2 and SPE test statistics can be employed

$$T^2 = \mathbf{x}^T \mathbf{R} (\frac{\mathbf{T}^T \mathbf{T}}{N-1})^{-1} \mathbf{R}^T \mathbf{x}$$
$$SPE = \|\tilde{\mathbf{x}}\|, \tag{2.33}$$

with the thresholds defined by

$$J_{th}^{T^2} = \frac{\gamma(N^2 - 1)}{N(N - \gamma)} F_\alpha(\gamma, N - \gamma)$$
$$J_{th}^{SPE} = g\chi_\alpha^2(h), \tag{2.34}$$

where $g = S/2\mu$ and $h = 2\mu^2/S$ with μ and S as sample mean and variance of SPE statistic [87].

Same as PCA, PLS method is suitable for the cases where the measurements follow multivariate Gaussian distribution. For dynamic processes where serial correlation appears in measurements, the so-called dynamic PLS (DPLS) approach has been developed in the same way as DPCA [66, 72, 91].

Recently, in [139] it has been revealed that the classical PLS approach may result in variation in $\hat{X}$ orthogonal to Y and $\tilde{X}$ may contain large variability of X and is therefore not suitable to monitor the faults which affect the quality variables and leads to miss-classification of faults. Moreover the iterative procedure involved in computation of the regression model leads to difficulties to interpret the PLS model. To solve the above mentioned problems a new modified approach has been proposed in [128], which avoids the above mentioned drawbacks and is computationally more simple.

2.4 Subspace-based design of FDD systems

So far the bases of model-based FDD techniques and multivariate statistical process monitoring methods have been discussed. Model-based techniques rely on the rigorous development of process model and design of observer or parity space for FDD purpose. On the other hand, SPM methods involve development of monitoring scheme based on the historical process data. Since the process modeling based on first principles cannot be achieved in all cases, from the application point of view it would be interesting to obtain process model from historical data and then utilize the classical model-based techniques to design an FDD system.

Recent developments in linear algebra and statistics provide the possibility to identify the state space model of system in Eq. (2.2) directly from historical data. These techniques are known as subspace identification methods (SIM) [107]. Due to their numerical robustness, convergence properties, use of well-developed algorithm and the ability to deal with large amount of data, their application in this field is becoming vaster.

In SIM the state space model of linear time invariant dynamic process:

$$\mathbf{x}(k+1) = \mathbf{A}\mathbf{x}(k) + \mathbf{B}\mathbf{u}(k) + \mathbf{w}(k)$$
$$\mathbf{y}(k) = \mathbf{C}\mathbf{x}(k) + \mathbf{D}\mathbf{u}(k) + \mathbf{v}(k) \qquad (2.35)$$

with

$$E\left\{ \begin{bmatrix} \mathbf{w}(i) \\ \mathbf{v}(i) \end{bmatrix} \begin{bmatrix} \mathbf{w}^T(j) & \mathbf{v}^T(j) \end{bmatrix} \right\} = \begin{bmatrix} \mathbf{Q} & \mathbf{S} \\ \mathbf{S}^T & \mathbf{R} \end{bmatrix} \delta_{ij} \geq 0$$

is directly identified from given input and output data through a projection of the row space of certain block Hankel matrices of data into the row spaces of other block Hankel matrices [107]. The matrices $\mathbf{Q} \in \mathbb{R}^{n \times n}$, $\mathbf{S} \in \mathbb{R}^{n \times m}$ and $\mathbf{R} \in \mathbb{R}^{m \times m}$ are covariance matrices of the noise sequences $\mathbf{w}(k)$ and $\mathbf{v}(k)$. The most commonly used SIMs are numerical subspace state space system identification (N4SID; [106]), multivariable output error state space (MOESP; [113, 114, 112]) and canonical variate analysis (CVA; [74]).

The key step in SIM is to identify the system state and the extended observability matrix

$$\mathbf{\Gamma}_i = \begin{bmatrix} \mathbf{C}^T & (\mathbf{C}\mathbf{A})^T & (\mathbf{C}\mathbf{A}^2)^T & \cdots & (\mathbf{C}\mathbf{A}^{i-1})^T \end{bmatrix}^T$$

from input and output data and then calculate the system matrices by solving a least squares problem [78, 107]. The SIM provides us an LTI model of process which can capture the dynamic behavior of the system. The system matrices $\mathbf{A}, \mathbf{B}, \mathbf{C}$ and $\mathbf{D}$ can be used to design diagnostic observer as well as parity space-based residual generator as discussed in Sections 2.2.2 and 2.2.3 .

Recently in [32], the authors have proposed a new approach for data-driven design of fault detection and isolation (FDI) systems, which can be used for parity space and observer-based FDI as well as soft sensor construction. The basic idea is to directly identify the parity vector from historical data and then use it for parity space-based residual generation Eq. (2.18) or use its one-to-one relation with diagnostic observer described in Section 2.2.4 to build an observer-based residual generator. The procedure is summarized in Algorithm 1.

Algorithm 1. *Design of PS-based residual generator*

Step 1 *Generate data matrices* $\mathbf{Z}_f$ *and* $\mathbf{Z}_p$ *and construct* $\frac{1}{N}\mathbf{Z}_f\mathbf{Z}_p^T$ *as follows*

$$\mathbf{U}(j) = \begin{bmatrix} \mathbf{u}(j) & \mathbf{u}(j+1) & \cdots & \mathbf{u}(j+N-1) \end{bmatrix} \in \mathbb{R}^{l\times N}$$

$$\mathbf{Y}(j) = \begin{bmatrix} \mathbf{y}(j) & \mathbf{y}(j+1) & \cdots & \mathbf{y}(j+N-1) \end{bmatrix} \in \mathbb{R}^{m\times N}$$

$$\mathbf{U}_p = \begin{bmatrix} \mathbf{U}(i-s) \\ \vdots \\ \mathbf{U}(i-1) \end{bmatrix} \in \mathbb{R}^{sl\times N}, \ \mathbf{Y}_p = \begin{bmatrix} \mathbf{Y}(i-s) \\ \vdots \\ \mathbf{Y}(i-1) \end{bmatrix} \in \mathbb{R}^{sm\times N},$$

$$\mathbf{U}_f = \begin{bmatrix} \mathbf{U}(i) \\ \vdots \\ \mathbf{U}(i+s-1) \end{bmatrix} \in \mathbb{R}^{sl\times N}, \ \mathbf{Y}_f = \begin{bmatrix} \mathbf{Y}(i) \\ \vdots \\ \mathbf{Y}(i+s-1) \end{bmatrix} \in \mathbb{R}^{sm\times N},$$

$$\mathbf{Z}_p = \begin{bmatrix} \mathbf{Y}_p \\ \mathbf{U}_p \end{bmatrix} \in \mathbb{R}^{(sl+sm)\times N}, \ \mathbf{Z}_f = \begin{bmatrix} \mathbf{Y}_f \\ \mathbf{U}_f \end{bmatrix} \in \mathbb{R}^{(sl+sm)\times N} \tag{2.36}$$

Step 2 *Do an SVD on* $\frac{1}{N}\mathbf{Z}_f\mathbf{Z}_p^T$

$$\frac{1}{N}\mathbf{Z}_f\mathbf{Z}_p^T = \mathbf{U}\begin{bmatrix} \boldsymbol{\Sigma}_1 & 0 \\ 0 & \boldsymbol{\Sigma}_2 \end{bmatrix}\mathbf{V}^T, \quad \boldsymbol{\Sigma}_2 \in \mathbb{R}^{(sm-n)\times(sm-n)} \tag{2.37}$$

with unitary matrices $\mathbf{V}^T \in \mathbb{R}^{s(l+m)\times s(l+m)}$ *and*

$$\mathbf{U} = \begin{bmatrix} \mathbf{U}_{11} & \mathbf{U}_{12} \\ \mathbf{U}_{21} & \mathbf{U}_{22} \end{bmatrix} \in \mathbb{R}^{s(l+m)\times s(l+m)} \tag{2.38}$$

Step 3 *Set*

$$\boldsymbol{\Gamma}_s^{\perp} = \mathbf{U}_{12}^T \in \mathbb{R}^{(sm-n)\times sm}$$

$$\boldsymbol{\Gamma}_s^{\perp}\mathbf{H}_{s,u} = -\mathbf{U}_{22}^T \in \mathbb{R}^{(sm-n)\times sl} \tag{2.39}$$

Step 4 *Select* $\mathbf{v}_s$ *and* $\mathbf{v}_s\mathbf{H}_{s,u}$ *as two row vectors satisfying* $\mathbf{v}_s \in \boldsymbol{\Gamma}_s^{\perp}$ *and* $\mathbf{v}_s\mathbf{H}_{s,u} \in \boldsymbol{\Gamma}_s^{\perp}\mathbf{H}_{s,u}$

It is worth pointing out, to ensure that $\boldsymbol{\Gamma}_s^{\perp}$ and $\boldsymbol{\Gamma}_s^{\perp}\mathbf{H}_{s,u}$ are identified correctly, the input excitation condition should be satisfied [115]. Moreover, in [32], the authors have proposed an observer based residual generator using the one-to-one relationship. In [125], this method has been extended for monitoring of batch process.

The classical SIM requires high computation effort due to first identification of system states and observability matrix and then, based on it calculating the system matrices in the second step, while the method proposed in [32] is computationally much more simple. A comprehensive comparison study between popular data-driven process monitoring methods from implementation and application perspectives is provided in [126].

2.5 Concluding remarks

In this introductory chapter, representations of technical system are described. The disturbances and faults are included in the process model. The rest of the chapter is dedicated to the overview of the modern FDD techniques and their recent developments and extensions. The popular model-based FDD techniques, fault detection filter, diagnostic observer and parity space-based residual generator are elaborated and their interconnections are described.

Multivariate statistical process monitoring methods are introduced, their important concepts are described and their advantages and shortcomings are explained, with the help of principal component analysis and partial least squares. Moreover, an alternative solution which combines model-based and data-driven methods based on subspace identification techniques and its major developments in recent years are discussed. Their major drawbacks which limits their application to linear time-invariant systems are discussed in the forthcoming chapters and solutions are proposed.

3 Fault detection in multimode nonlinear systems

Modern complex industrial plants are commonly working on multiple operating conditions because of different product specifications and some external restrictions. As a consequence, the plant characteristics change from one operating condition to another due to nonlinearity and set-point changes in the system. Hence, the statistical model obtained from traditional MSPM techniques for an operating mode is not valid anymore for the others and will induce false alarms. This is because the basic assumption in MSPM methods that the data should follow uni-modal Gaussian distribution. Under the assumption that the data corresponding to each operating point follows a multivariate Gaussian distribution with various statistical properties, the available historical data can be seen as a mixture of Gaussian components with different mean vectors and covariance matrices. As an example the data obtained from a continuous stirred tank heater (CSTH) benchmark is shown in Fig. 3.1. An intuitive engineering approach to solve this problem is extracting labeled data for each operating mode to derive the statistical models and in on-line monitoring step use different statistical models for corresponding operating mode. This approach requires a *scheduling variable* which represent the actual mode of system and entails high engineering efforts. Moreover, those labeled data are not available always in practice.

Recently, some research efforts have been done for data-driven nonlinear process identification and monitoring based on the multiple model assumption with the help of mixture modeling tools. The PCA process monitoring method has been extended to be used for multimode process monitoring under Gaussian mixture model (GMM)

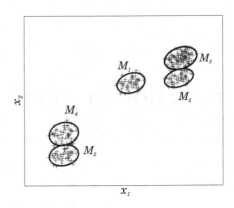

Figure 3.1: Scatter plot of data for CSTH plant under different operating points

assumption (see for example [135, 132, 15, 39, 40]) and their applications in monitoring of batch processes and semiconductor technology
have been reported (see for instance [17, 136, 133, 130, 134]). Nevertheless, their applications for performance-based fault detection and
quality monitoring have not been studied extensively. In [129] a two
step method has been developed for quality prediction in multi-phase
batch processes using DPLS method. In the first step the data are
classified using GMM method and in the second step the multi-phase
PLS method is employed to find the relation between the process
measurements and quality variable in each phase. Although, the effectiveness of the proposed method has been shown, the computation cost
of this approach is high due to complexity of classical PLS algorithm
and will grow by increasing the number of phases.

Motivated by the above observations, in this chapter a new approach
is proposed for monitoring the product quality in nonlinear systems
and detection of the quality related faults. The nonlinear system
is assumed to be linear around the operating points and therefore
considered as a piecewise linear system corresponding to each operating
mode. The data in each operating condition follows multivariate
normal distribution. Using a mixture modeling method, the regression
model which describes the correlation structure of plant measurements

and the quality variables is identified for each operating mode. Based on the regression coefficient the monitoring scheme is designed and using a Bayesian inference strategy a new index is used for fault detection purpose.

3.1 Preliminaries and problem formulation

PLS [48, 50] is recognized as a powerful tool to model the processes and discover their underlying structure and is used in numerous applications in chemometrics and industrial process monitoring. The goal of PLS regression is to predict the output data set (quality variables) $\mathbf{Y}$ from an input data set (process measurements) $\mathbf{X}$ and to describe their common structure. Specifically, PLS regression looks for a set of components (called latent variables) that performs a simultaneous decomposition of $\mathbf{X}$ and $\mathbf{Y}$ with the constraint that these components explain the covariance between $\mathbf{X}$ and $\mathbf{Y}$ as much as possible. The correlation model of process measurements and product quality which is identified by PLS algorithm is shown in Eq. (2.32) and the standard PLS algorithm, known as NIPALS is described in Section 2.3.2. Moreover, some shortcomings of the traditional PLS approach is explained therein.

The application of standard PLS algorithm is based on the unimodel multivariate normal distribution of $\mathbf{X}$ and $\mathbf{Y}$. However, in some industrial applications, the process has nonlinear structure and standard PLS approach cannot be used to model the underlying structure due to operating point changes.

To cope with this problem, in this chapter a new approach is presented which avoids the computational drawback of standard PLS algorithm and provides a complete decomposition of measurements space with respect to quality variables. Based on that, a novel solution for quality monitoring in nonlinear systems is proposed, assuming that the model of nonlinear processes can be represented with finite number of linear models corresponding to each operating point. Moreover, it is assumed that the data for each linear model follows multivariate

Gaussian distribution. Thus, the training data available for design of monitoring scheme is a *mixture of finite Gaussian components* [82].

3.2 Modified PLS approach

It has been mentioned in Section 2.3.2 that the classical PLS algorithm involves iterative solution of an optimization problem, which makes it difficult to interpret the PLS model. Moreover, it has been shown in [139] that PLS algorithm entails an oblique projection. Thus, the measurements space is not correctly decomposed according to its correlation to the quality variable. To this aim, in [128] a new modified approach has been developed which is computationally simple and easy to interpret. Using this approach, when $N \gg max(n, m)$, Eq. (2.32) can be written as follows

$$\frac{1}{N-1}\mathbf{Y}^T\mathbf{X} = \frac{1}{N-1}\mathbf{M}^T\mathbf{X}^T\mathbf{X} + \frac{1}{N-1}\mathbf{E}_y^T\mathbf{X} \approx \mathbf{M}^T\frac{\mathbf{X}^T\mathbf{X}}{N-1}. \quad (3.1)$$

This is due to the fact that $\text{cov}(\mathbf{e}_y, \mathbf{x}) = \mathbf{0}$. Thus, the regression coefficient can be calculated as

$$\mathbf{M} = (\mathbf{X}^T\mathbf{X})^\dagger\mathbf{X}^T\mathbf{Y}, \quad (3.2)$$

where $(.)^\dagger$ represent the pseudo-inverse. Based on it, $\mathbf{X}$ can be decomposed into two parts, $\hat{\mathbf{X}}$ and $\tilde{\mathbf{X}}$ such that $\hat{\mathbf{X}}$ is fully correlated to $\mathbf{Y}$ and $\tilde{\mathbf{X}}$ is orthogonal to $\hat{\mathbf{X}}$ and has no contribution in predicting $\mathbf{Y}$:

$$\mathbf{X} = \hat{\mathbf{X}} + \tilde{\mathbf{X}}$$
$$\mathbf{Y} = \mathbf{X}\mathbf{M} + \mathbf{E}_y \quad (3.3)$$

To detect the faults which effects the product quality $\mathbf{Y}$, an index is proposed based on monitoring the $\hat{\mathbf{X}}$ subspace

$$T_{\hat{x}}^2 = \mathbf{x}^T\mathbf{P}_M(\frac{\mathbf{P}_M^T\mathbf{X}^T\mathbf{X}\mathbf{P}_M}{N-1})^{-1}\mathbf{P}_M^T\mathbf{x}, \quad (3.4)$$

where $\mathbf{P}_M \in \mathbb{R}^{n \times m}$ is obtained by performing SVD on $\mathbf{MM}^T$

$$\mathbf{MM}^T = \begin{bmatrix} \mathbf{P}_M & \tilde{\mathbf{P}}_M \end{bmatrix} \begin{bmatrix} \mathbf{\Lambda}_M & 0 \\ 0 & 0 \end{bmatrix} \begin{bmatrix} \mathbf{P}_M^T \\ \tilde{\mathbf{P}}_M^T \end{bmatrix} \tag{3.5}$$

and the threshold for fault detection follows

$$J_{th}^{T_{\hat{x}}^2} = \frac{m(N^2 - 1)}{N(N - m)} F_\alpha(m, N - m), \tag{3.6}$$

where $F_\alpha(m, N - m)$ is F-distribution with parameters m and $N - m$ and α is the confidence level. If the number of training samples N is large enough then

$$J_{th}^{T_{\hat{x}}^2} = \chi_\alpha^2(m), \tag{3.7}$$

where $\chi_\alpha^2(m)$ is χ^2 distribution with m degrees of freedom and confidence level α.

Same as $\hat{\mathbf{X}}$, the subspace $\tilde{\mathbf{X}}$ can be monitored to detect the faults in the system which are not affecting the product quality variables $\mathbf{Y}$ using

$$T_{\tilde{x}}^2 = \mathbf{x}^T \tilde{\mathbf{P}}_M \left(\frac{\tilde{\mathbf{P}}_M^T \mathbf{X}^T \mathbf{X} \tilde{\mathbf{P}}_M}{N - 1} \right)^{-1} \tilde{\mathbf{P}}_M^T \mathbf{x}, \tag{3.8}$$

with the threshold

$$J_{th}^{T_{\tilde{x}}^2} = \frac{(n - m)(N^2 - 1)}{N(N - n + m)} F_\alpha(n - m, N - n + m). \tag{3.9}$$

To monitor the subspace $\mathbf{E}_y$ in Eq. (2.32) which is not correlated to $\mathbf{X}$, an SPE index can be employed

$$SPE_y = ||\mathbf{y} - \mathbf{M}^T \mathbf{x}||^2 \tag{3.10}$$

with the corresponding threshold

$$J_{th}^{SPE_y} = g\chi_\alpha^2(h_y), \tag{3.11}$$

where $g = S/2\mu$ and $h_y = 2\mu^2/S$ and μ, S are the mean and variance of SPE_y respectively.

Moreover, the matrix $\mathbf{M}$ identified by Eq. (3.2) can be used for prediction of quality variable using Eq. (2.32). It can bee seen that this approach performs an orthogonal decomposition of variables space and therefore decomposes it into two subspaces, one is responsible for variation of quality variable and the other has no contribution to it. From the computational point of view it entails maximum two times SVD compared to classical method which requires γ times SVD, where γ is the number of latent variables. Furthermore, in classical PLS approach the number of latent variables should be specified in advance, while this new algorithm does not rely on that.

3.3 Multimode process monitoring

Consider a nonlinear process working in K different operating modes $\mathcal{M}_1, \mathcal{M}_2, \cdots, \mathcal{M}_K$, in which, each mode is characterized by Eq. (3.3) with different model parameters $\mathbf{M}_i$, $i = 1, \cdots, K$. The purpose is to design a monitoring scheme for the process using the modified PLS approach in Eqs. (3.1) to (3.11) assuming that the model parameters for each operating mode are unknown [44].

3.3.1 Estimation of model parameters

To design the monitoring scheme, first the mixture model should be identified, using historical data. The historical data set contains measurements for all normal process operating modes. Assume that the available historical data $\mathcal{D}$ is collected from N different samples, each sample contains the measurements of process $\mathbf{x} \in \mathbb{R}^l$ and quality variables $\mathbf{y} \in \mathbb{R}^m$:

$$\mathcal{D} = \left\{ \begin{bmatrix} \mathbf{y}_1 \\ \mathbf{x}_1 \end{bmatrix}, \begin{bmatrix} \mathbf{y}_2 \\ \mathbf{x}_2 \end{bmatrix}, \cdots, \begin{bmatrix} \mathbf{y}_N \\ \mathbf{x}_N \end{bmatrix} \right\}$$
$$= \{\mathbf{d}_1, \mathbf{d}_2, \cdots, \mathbf{d}_N\}, \tag{3.12}$$

where $\mathbf{d}_k \in \mathbb{R}^{l+m}$ for $k = 1, \cdots, N$ is a sample from a multimode process. The probability density function (PDF) of an arbitrary sample $\mathbf{d}$ from the multimode static process can be represented by a finite Gaussian mixture model (FGMM) which is a weighted sum of multiple local Gaussian components:

$$p(\mathbf{d}|\theta) = \sum_{i=1}^{K} w_i g(\mathbf{d}|\theta_i). \tag{3.13}$$

K is the number of mixture components, w_i is the weight of ith component $\mathcal{M}_i$, θ_i are sets of parameters of the ith Gaussian components,

$$\theta_i = \{w_i, \boldsymbol{\mu}_{x,i}, \boldsymbol{\mu}_{y,i}, \boldsymbol{\Sigma}_{xx,i}, \boldsymbol{\Sigma}_{xy,i}, \boldsymbol{\Sigma}_{yy,i}\} \tag{3.14}$$

and $g(\mathbf{d}|\theta_i)$ is corresponding multivariate Gaussian density function for the component $\mathcal{M}_i$,

$$g(\mathbf{d}|\theta_i) = \frac{1}{(2\pi)^{m/2}|\boldsymbol{\Sigma}_i|^{1/2}} \exp\left(-\frac{1}{2}(\mathbf{d} - \boldsymbol{\mu}_i)^T \boldsymbol{\Sigma}_i^{-1}(\mathbf{d} - \boldsymbol{\mu}_i)\right),$$

with

$$\boldsymbol{\mu}_i = \begin{bmatrix} \boldsymbol{\mu}_{x,i} \\ \boldsymbol{\mu}_{y,i} \end{bmatrix}, \quad \boldsymbol{\Sigma}_i = \begin{bmatrix} \boldsymbol{\Sigma}_{xx,i} & \boldsymbol{\Sigma}_{xy,i} \\ \boldsymbol{\Sigma}_{xy,i}^T & \boldsymbol{\Sigma}_{yy,i} \end{bmatrix}.$$

It is worth pointing out that $\sum_{i=1}^{K} w_i = 1$ with $0 \leq w_i \leq 1$.

To construct the monitoring scheme, the following set of parameters should be identified

$$\Theta = \{\theta_1, \theta_2, \cdots, \theta_K\} \tag{3.15}$$

This can be done by assignment of log-likelihood of mixture components as

$$\log p(\mathcal{D}|\Theta) = \log \prod_{k=1}^{N} p(\mathbf{d}_k|\Theta) = \sum_{k=1}^{N} \log \sum_{i=1}^{K} w_i p(\mathbf{d}_k|\theta_i). \tag{3.16}$$

The maximum likelihood estimate (MLE) can be achieved by

$$\hat{\Theta}_{MLE} = \arg\max_{\Theta}\{\log p(\mathcal{D}|\Theta)\}, \qquad (3.17)$$

or alternatively the solution can be achieved by maximum a posteriori (MAP) criterion

$$\hat{\Theta}_{MAP} = \arg\max_{\Theta}\{\log p(\mathcal{D}|\Theta) + \log p(\Theta)\}. \qquad (3.18)$$

The solution of Eq. (3.17) and Eq. (3.18) cannot be found analytically. The Expectation-Maximization (EM) algorithm can be used for this purpose [81]. EM is an iterative algorithm which finds the local maxima of log-likelihood functions in Eq. (3.17) or Eq. (3.18). The EM algorithm is based on the assumption that $\mathcal{D}$ is an incomplete data set in which the missing part in finite mixture modeling can be interpreted as N tags, $\mathcal{Z} = \{z_1, \cdots, z_N\}$ represents each sample generated in the respective operating mode. In EM algorithm the conditional expectation of the log-likelihood for complete data $\mathcal{C} = \{\mathcal{D}, \mathcal{Z}\}$ is calculated in E-step as follows

$$\mathcal{Q}(\Theta|\Theta^{old}) = E\{\log p(\mathcal{D}, \mathcal{Z}|\Theta)|\mathcal{D}, \Theta^{old}\} \qquad (3.19)$$

and updates the estimation of parameters in M-step using

$$\Theta = \arg\max_{\Theta} \mathcal{Q}(\Theta|\Theta^{old}) \qquad (3.20)$$

for MLE or

$$\Theta = \arg\max_{\Theta} \mathcal{Q}(\Theta|\Theta^{old} + \log p(\Theta)) \qquad (3.21)$$

based on MAP estimation. For more information about EM algorithm and its derivation the readers are referred to Appendix A.

To estimate the set of parameters in Eq. (3.15), the conditional expectation in Eq. (3.19) can be written as shown in Eq. (3.22).

$$
\begin{aligned}
\mathcal{Q}(\Theta|\Theta^{old}) &= E\{\log p(\mathbf{y}_N, \cdots, \mathbf{y}_1, \mathbf{x}_N, \cdots, \mathbf{x}_1, z_N, \cdots, z_1|\Theta)|\mathcal{D}, \Theta^{old}\} \\
&= E\{\log p(\mathbf{y}_N, \cdots, \mathbf{y}_1|\mathbf{x}_N, \cdots, \mathbf{x}_1, z_N, \cdots, z_1, \Theta) \times \\
&\qquad p(\mathbf{x}_N, \cdots, \mathbf{x}_1|z_N, \cdots, z_1, \Theta)p(z_N, \cdots, z_1|\Theta)|\mathcal{D}, \Theta^{old}\} \\
&= E\{\sum_{k=1}^{N} \log p(\mathbf{y}_k|\mathbf{x}_k, z_k, \Theta) + \\
&\qquad \log p(\mathbf{x}_k|z_k, \Theta) + \log p(z_k|\Theta)|\mathcal{D}, \Theta^{old}\} \\
&= \sum_{k=1}^{N}\sum_{i=1}^{K} p(z_k = i|\mathcal{D}, \Theta^{old}) \log p(\mathbf{y}_k|\mathbf{x}_k, \theta_i) + \\
&\quad \sum_{k=1}^{N}\sum_{i=1}^{K} p(z_k = i|\mathcal{D}, \Theta^{old}) \log p(\mathbf{x}_k|\theta_i) + \\
&\quad \sum_{k=1}^{N}\sum_{i=1}^{K} p(z_k = i|\mathcal{D}, \Theta^{old}) \log p(z_k = i|\theta_i) \qquad (3.22)
\end{aligned}
$$

where Θ^{old} are the unknown parameters obtained in the previous iteration.

In derivation of Eq. (3.22), it is assumed that the quality variable at k^{th} sampling time instant is independent of past value of missing variable z and process variables $\mathbf{x}$ and only depends on their current values. The same assumption is also made for process variables $\mathbf{x}$. Thus

$$
p(\mathbf{y}_N, \cdots, \mathbf{y}_1|\mathbf{x}_N, \cdots, \mathbf{x}_1, z_N, \cdots, z_1, \Theta) = \prod_{k=1}^{N} p(\mathbf{y}_k|\mathbf{x}_k, z_k, \Theta)
$$

$$
p(\mathbf{x}_N, \cdots, \mathbf{x}_1|z_N, \cdots, z_1, \Theta) = \prod_{k=1}^{N} p(\mathbf{x}_k|z_k, \Theta) \qquad (3.23)
$$

These assumptions are valid, since it has been assumed that there is no dynamic behavior in the system and the quality variable at each instant depends only on temporal data and current mode of system.

Moreover the conditional distributions of the variables in the regression model in Eq. (3.3) are as follows

$$\mathbf{x}_k|z_k = i, \Theta \sim \mathcal{N}(\boldsymbol{\mu}_{x,i}, \boldsymbol{\Sigma}_{xx,i})$$
$$\mathbf{y}_k|\mathbf{x}_k, z_k = i, \Theta \sim \mathcal{N}(\boldsymbol{\mu}_{y|x,i}, \boldsymbol{\Sigma}_{y|x,i}) \tag{3.24}$$

where

$$\boldsymbol{\mu}_{y|x,i} = \boldsymbol{\mu}_{y,i} + \boldsymbol{\Sigma}_{xy,i}^T \boldsymbol{\Sigma}_{xx,i}^{-1}(\mathbf{x}_k - \boldsymbol{\mu}_{x,i})$$
$$\boldsymbol{\Sigma}_{y|x,i} = \boldsymbol{\Sigma}_{yy,i} - \boldsymbol{\Sigma}_{xy,i}^T \boldsymbol{\Sigma}_{xx,i}^{-1} \boldsymbol{\Sigma}_{xy,i}. \tag{3.25}$$

The M-step in EM algorithm is carried out by derivation of the conditional expectation, $\mathcal{Q}(\Theta|\Theta^{old})$, with respect to the relevant unknown parameters. After performing the derivation and equating it to zero, the updates of parameters in M-step are as follows:

$$\boldsymbol{\mu}_{x,i} = \frac{\sum\limits_{k=1}^{N} p(\mathcal{M}_i|\mathbf{d}_k)\mathbf{x}_k}{\sum\limits_{k=1}^{N} p(\mathcal{M}_i|\mathbf{d}_k)}, \quad \boldsymbol{\mu}_{y,i} = \frac{\sum\limits_{k=1}^{N} p(\mathcal{M}_i|\mathbf{d}_k)\mathbf{y}_k}{\sum\limits_{k=1}^{N} p(\mathcal{M}_i|\mathbf{d}_k)}$$

$$\boldsymbol{\Sigma}_{xx,i} = \frac{\sum\limits_{k=1}^{N} p(\mathcal{M}_i|\mathbf{d}_k)(\mathbf{x}_k - \boldsymbol{\mu}_{x,i})(\mathbf{x}_k - \boldsymbol{\mu}_{x,i})^T}{\sum\limits_{k=1}^{N} p(\mathcal{M}_i|\mathbf{d}_k)}$$

$$\boldsymbol{\Sigma}_{yy,i} = \frac{\sum\limits_{k=1}^{N} p(\mathcal{M}_i|\mathbf{d}_k)(\mathbf{y}_k - \boldsymbol{\mu}_{y,i})(\mathbf{y}_k - \boldsymbol{\mu}_{y,i})^T}{\sum\limits_{k=1}^{N} p(\mathcal{M}_i|\mathbf{d}_k)}$$

$$\boldsymbol{\Sigma}_{xy,i} = \frac{\sum\limits_{k=1}^{N} p(\mathcal{M}_i|\mathbf{d}_k)(\mathbf{x}_k - \boldsymbol{\mu}_{x,i})(\mathbf{y}_k - \boldsymbol{\mu}_{y,i})^T}{\sum\limits_{k=1}^{N} p(\mathcal{M}_i|\mathbf{d}_k)}$$

$$w_i = \frac{\sum\limits_{k=1}^{N} p(\mathcal{M}_i|\mathbf{d}_k)}{N} \qquad (3.26)$$

where $p(\mathcal{M}_i|\mathbf{d}_k)$ is calculated in E-step using Bayes' rule as

$$p(\mathcal{M}_i|\mathbf{d}_k) = \frac{w_i g(\mathbf{d}_k|\Theta_i)}{\sum\limits_{i=1}^{K} w_i g(\mathbf{d}_k|\Theta_i)}. \qquad (3.27)$$

After the parameter estimation using EM algorithm the regression coefficient in PLS model can be calculated in the same way as shown in Eq. (3.2):

$$\mathbf{M}_i = \Sigma_{xx,i}^{-1} \Sigma_{xy,i} \qquad (3.28)$$

or in the case that $\Sigma_{xx,i}$ is not a full rank matrix

$$\mathbf{M}_i = \Sigma_{xx,i}^{\dagger} \Sigma_{xy,i}. \qquad (3.29)$$

In next step the estimated parameters $\mathbf{M}_i, \boldsymbol{\mu}_{x,i}, \boldsymbol{\mu}_{y,i}, \Sigma_{xx,i}, \Sigma_{yy,i}$ will be used to design the monitoring scheme. The focus will be on the detection of faults, which influence the quality of products.

3.3.2 Design of monitoring scheme

It has been mentioned earlier in Section 2.3.2 that the main application of PLS method is to detect the quality related faults in the process when the on-line measurement for product quality is not available. The on-line process measurements at time instant k, $\mathbf{x}(k)$, are used together with the PLS model obtained in off-line design step to detect the faults in the system. Similarly here, for the fault detection purpose an index is further defined to represent the probability that the monitored sample, $\mathbf{x}(k)$ belongs to a fault

$$J_g(k) = p(\mathbf{x}(k) \in f). \qquad (3.30)$$

This index can be obtained through marginalization as

$$J_g(k) = \sum_{i=1}^{K} p(\mathbf{x}(k) \in f | \mathbf{x}(k) \in \mathcal{M}_i) p(\mathbf{x}(k) \in \mathcal{M}_i). \qquad (3.31)$$

The second term on the right side of Eq. (3.31) represents the probability that the given sample belongs to component $\mathcal{M}_i$ which can be calculated from the PDF of multivariate normal distribution given the estimated parameters for each mode and a Bayesian inference strategy [135]

$$
\begin{aligned}
p(\mathbf{x}(k) \in \mathcal{M}_i) &= \frac{p(\mathbf{x}(k)|\mathcal{M}_i)p(\mathcal{M}_i)}{p(\mathbf{x}(k))} \\
&= \frac{p(\mathbf{x}(k)|\mathcal{M}_i)p(\mathcal{M}_i)}{\sum\limits_{i=1}^{K} p(\mathbf{x}(k)|\mathcal{M}_i)p(\mathcal{M}_i)} \\
&= \frac{w_i g(\mathbf{x}(k)|\boldsymbol{\mu}_{x,i}, \boldsymbol{\Sigma}_{xx,i})}{\sum\limits_{i=1}^{K} w_i g(\mathbf{x}(k)|\boldsymbol{\mu}_{x,i}, \boldsymbol{\Sigma}_{xx,i})}.
\end{aligned}
\qquad (3.32)
$$

The first term on the right hand side of Eq. (3.31) represents the probability of fault provided that the sample belongs to component $\mathcal{M}_i$. To calculate $p(\mathbf{x}(k) \in f | \mathbf{x}(k) \in \mathcal{M}_i)$, the $T_{\hat{x}}^2$ index introduced in Eq. (3.4) will be utilized. This probability can be written as

$$p(\mathbf{x}(k) \in f | \mathbf{x}(k) \in \mathcal{M}_i) = p(T_{\hat{x}}^2(\mathbf{x}, i) \leq T_{\hat{x}}^2(\mathbf{x}(k), i)), \qquad (3.33)$$

which can be calculated by integrating the F probability density function in (3.6) or χ^2 distribution in (3.7) with appropriate degrees of freedom. The right hand side of Eq. (3.33) can be interpreted as the probability that the calculated T^2 index corresponding to measurement sample $\mathbf{x}(k)$ (namely $T_{\hat{x}}^2(\mathbf{x}(k), i)$) is greater than or equal to the obtained T^2 index from off-line fault free data ($T_{\hat{x}}^2(\mathbf{x}, i)$), assuming that the data are generated in mode $\mathcal{M}_i$. Furthermore, since $0 \leq J_g(k) \leq 1$ a confidence level $(1 - \alpha)$ can be specified for fault detection purpose with the hypothesis as follows

$$\begin{cases} J_g \leq 1 - \alpha & \text{fault free} \\ J_g > 1 - \alpha & \text{faulty.} \end{cases} \tag{3.34}$$

The procedure for design of the proposed monitoring scheme is shown in Algorithm 2.

Algorithm 2. *Design of fault detection system*

Step 1 *Collect the normal operation data from different operating modes.*

Step 2 *Apply the EM algorithm to estimate the parameters shown in (3.14) using (3.26) and (3.27).*

Step 3 *Obtain* $\mathbf{M}_i$ *using (3.28) or (3.29) for* $i = 1, \cdots K$.

Step 4 *In on-line step when a new sample of measurements is available, perform following steps for* $i = 1, \cdots, K$:

4.1 *Compute* $T_{\hat{x}}^2(\mathbf{x}(k), i)$ *using (3.4) and the identified parameters of* $\mathcal{M}_i$.

4.2 *Compute* $p(\mathbf{x}(k) \in \mathcal{M}_i)$ *using the multivariate Gaussian PDF and Eq. (3.32).*

4.3 *Compute* $p(\mathbf{x}(k) \in f | \mathbf{x}(k) \in \mathcal{M}_i)$ *using* $T_{\hat{x}}^2(\mathbf{x}(k), i)$ *and (3.33).*

Step 5 *Calculate the fault detection index in (3.31).*

Step 6 *Use the fault detection hypothesis in (3.34) to detect fault and go to step 4.*

3.4 An illustrative example

The method proposed in Section 3.3 is implemented on the CSTH Simulink benchmark proposed in [103]. The description of the benchmark and its diagram is given in Appendix B and Fig. B.1, respectively. CSTH plants are widely used in industry. Inside the tank a chemical reaction takes place under certain temperature and level. The main objective of the control system is to maintain the temperature and level constant according to reaction specification. For this simulation the temperature is considered as the quality variable, hence it strongly affects the final quality of the product. The plant's normal operating conditions are specified by the set points shown in Table 3.1.

Table 3.1: Sets of normal operating modes

Set points	Mode 1	Mode 2	Mode 3
Level $[mA]$	12	12	12
Temperature $[mA]$	10.5	10.5	10
Hot water valve $[mA]$	0	5	4

The off-line training step is performed using the data obtained from these three operating modes according to steps 1-3 of Algorithm 2. For on-line monitoring step, 50 samples of measurements from each operating point are generated. In addition, two different fault scenarios are considered: f_1, a sensor fault in level and f_2, an actuator fault in steam valve. The plot of plant measurements for on-line monitoring is shown in Fig. 3.2 where the vertical dashed lines represent the samples where the modes are changed. The fault f_1 in level sensor happens from sample 100 to 150 and does not affect the product quality, namely temperature. The fault f_2 in hot water valve happens from sample 151 to 200 and causes a bias in temperature, as it can be seen from Fig. 3.2. There is another change in temperature from samples 201 to 250 which is due to product specification change and corresponds to the mode 3 of the operation. The on-line monitoring is performed according to steps 4-6 of Algorithm 2. The fault detection index shown in Eq. (3.31) is calculated and plotted in Fig. 3.3. The horizontal dashed line represents the threshold with 95% confidence level. It can be seen that the fault detection index is sensitive to the faults which are affecting the quality variable.

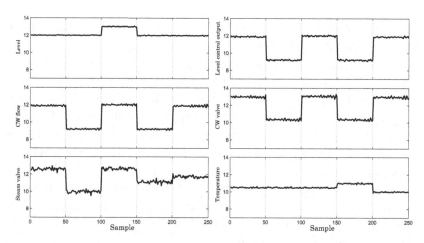

Figure 3.2: On-line process measurements

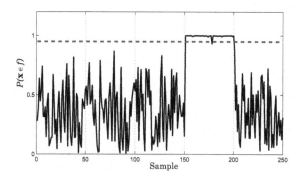

Figure 3.3: Fault detection indicator

3.5 Concluding remarks

In this chapter, a novel technique for design of fault detection system
in nonlinear multimode processes has been proposed. The focus of
the method is to detect those faults which affect the product quality
or the performance variables in the process. The nonlinear process
is assumed as a piecewise linear system and its correlation model
is identified with the help of EM algorithm. Furthermore, a unique

index is defined which represents the probability of a quality related fault happening in the system.

As one of its important features, this method does not require labeled data for off-line modeling. The classification is performed inherently in EM algorithm through the hidden variable, $\mathcal{Z}$. Moreover, this method does not rely on scheduling variable for on-line monitoring. Instead of that, the *a posterior* probabilities that the sample is generated in different modes are integrated into the detection scheme.

The application of the proposed scheme is restricted to the static processes. The concept will be extended in the next chapter to cover the fault detection problem in nonlinear dynamic multimode processes. Furthermore, this method will be used in the forthcoming chapters as a basis for fault isolation problem.

4 Fault detection in multimode nonlinear dynamic systems

Multivariate statistical process monitoring methods are powerful tools for detecting faults in industrial systems. However, industrial processes are often subjected to the dynamic changes. This dynamic behavior which is mainly due to set-point changes, causes mean shifts and covariance changes in the data. In Chapter 3 a new approach is presented which is able to deal with such mean and covariance changes. However, it cannot capture the serial correlation in data. Although this issue can be addressed by augmenting the past data with temporal data in the same manner as DPCA or DPLS [45], but the problems due to the transient behavior which often happens in the batch processes cannot be solved.

In general, the dynamic behavior of such processes is determined by interacting continuous and discrete dynamics where discrete dynamics arises due to local model changes, hence they are usually called hybrid systems in control community. In recent years, many research efforts have been devoted for identification, monitoring and control of hybrid systems and several methods have been developed.

In this chapter, first an overview of the available techniques for hybrid system identification is given. The comparison of the methods, their advantages and drawbacks are also explained. Then, a new approach is proposed for identification of the parity space representation of the multimode systems and based on it a novel scheme for quality monitoring in multimode nonlinear dynamic system is presented.

4.1 Hybrid system identification

Due to their vast application scope, hybrid systems have received increasing attention in recent years. Basically, hybrid systems are dynamical systems whose behavior can be interpreted as mixed continuous and discrete dynamics. Therefore, multimode dynamical processes can also be described in this category. The continuous dynamics in multimode systems is due to the dynamic behavior of each local model which can be described in state space form or other dynamical representation (see Section 2.1). The discrete dynamics appears in multimode systems when the operating point is changing and enables approximating the nonlinear dynamical system as a mixture of linear systems.

In hybrid system identification context, most of the research efforts have been dedicated to identification of piecewise affine (PWA) models. The identification task is summarized as classification, where each sample of data has to be associated to the most suitable local model, and parameter estimation where the parameters of each local model are obtained. An overview of different traditional techniques for identification of PWA systems is provided in [95].

Recently, some novel approaches to this problem have been proposed. In [93], a new sequential method has been developed to identify a piecewise autoregressive exogenous (PWARX) model in the form

$$y(k) = \phi^T(k)\boldsymbol{\theta}_i, \qquad (4.1)$$

where $\phi(k)$ represents the past input and output data and $\boldsymbol{\theta}_i$ is the parameters of the ith local model. To estimate the unknown parameter vector $\boldsymbol{\theta}_i$, an iterative algorithm has been developed which minimizes the error function

$$\Phi = \sum_{i=1}^{K} \sum_{k=1}^{N} (y(k) - y(k,i))^2 p(k,i)$$
$$y(k,i) = \phi^T(k)\boldsymbol{\theta}_i, \qquad (4.2)$$

where the $p(k,i)$ is the weight of sample k with respect to mode i. Each iteration in the proposed algorithm consists of two steps, where first the weightings $p(k, i)$ are determined and then the local ARX models are identified given the weights.

In [111], the authors have proposed a novel method to identify piecewise linear state space model using subspace identification method. The system description which has been considered in their work is in following form

$$\mathbf{x}(k+1) = \sum_{i=1}^{K} p(k, i) \left(\mathbf{A}_i \mathbf{x}(k) + \mathbf{B}_i \mathbf{u}(k) \right)$$

$$\mathbf{y}(k) = \sum_{i=1}^{K} p(k, i) \left(\mathbf{C}_i \mathbf{x}(k) + \mathbf{D}_i \mathbf{u}(k) \right). \qquad (4.3)$$

It has been assumed that the switching signal $p(k, i)$ is available. Moreover issues arising from state transformation in subspace identification have been addressed in this paper. The solution proposed in this paper has been extended for the case where switching signal is not available in [64]. An iterative solution based on subspace identification is given to identify the model parameters in Eq. (4.3) to determine the switching signal $p(k, i)$ and the number of modes K. To determine the switching signal, the method developed in [10] which detects the rank variations in the projected subspace has been used.

Statistical clustering methods based on GMM and support vector classifiers (SVC) have been used in [86] to estimate the parameters of PWARX model in Eq. (4.1). In this approach, GMM is used to classify the measured data and the boundary of hyperplane of two adjacent regions is determined using SVC. Least squares method is used further to identify the parameters of local models.

In [122, 123], a new approach has been developed for identification of the mixture of ARMA models for PWA systems and used for time series clustering purpose. The EM algorithm has been used to identify the model parameters and the number of local models has been determined using Bayesian information criterion (BIC).

A Bayesian approach for identification of hybrid system in PWARX form in Eq. (4.1) has been proposed in [65]. In this approach, the unknown model parameters are considered as random variables with specific distributions. The identification problem is treated as computing a posterior probability of the model parameters given the observation and prior knowledge.

In [60], a robust identification technique for PWARX model has been developed using EM technique and its robustness against outliers has been demonstrated. The method has been further extended to multiple model linear parameter varying (LPV) approach to deal the identification problem in nonlinear systems [62]. The scheduling variable which shows the transition between the local models has been integrated in the EM algorithm to determine the validity width of each local model in a nonlinear system. The extensions of these methods to PWARX system with Markovian switching dynamic and nonlinear parameter varying systems with missing observations, are given in [61] and [28], respectively.

Numerous methods have been developed to approach the identification problem in nonlinear systems using PWA assumption, nevertheless the data driven design of FD schemes for such systems has not been studied. To this end, in this chapter a new method has been developed for identification of parity space-based residual generator for multimode dynamic system considering PWA behavior of the system.

4.2 Preliminaries and problem formulation

In a real system which is working in different operating regimes, the model may change due to nonlinearities in the system and any deviation of model parameters will trigger a false alarm. To cope with this problem, updating the model repeatedly at each step could be a solution. But when the fault is growing slowly the model will be adapted to the fault and the residual will become insensitive to the fault [84, 85]. Another solution is to identify a model for each operating regime, design the FD system based on the identified models and combine the residual signals to detect the abnormal behavior of the plant. In our study, we propose a method to design a fault detection system based on the latter concept, using direct identification of the PS model. The PS-based residual generator has been introduced in Section 2.2.3 and its data-driven realization is elaborated in Section 2.4 and Algorithm 1, for LTI systems.

Consider a dynamic system which is working in K different modes $\mathcal{M}_1, \mathcal{M}_2, \cdots, \mathcal{M}_K$, in which the relation between the process measurements and product quality, in each operating mode, can be characterized by the state space representation

$$
\begin{aligned}
\mathbf{x}(k+1) &= \mathbf{A}_i \mathbf{x}(k) + \mathbf{B}_i \mathbf{u}(k) + \mathbf{w}(k) \\
y(k) &= \mathbf{C}_i \mathbf{x}(k) + \mathbf{D}_i \mathbf{u}(k) + v(k)
\end{aligned}
\tag{4.4}
$$

where $\mathbf{x} \in \mathbb{R}^n$ represents the states, $\mathbf{u} \in \mathbb{R}^l$ is a vector of process measurements and $y \in \mathbb{R}$ is the product quality measurement. The matrices $\mathbf{A}_i$, $\mathbf{B}_i$, $\mathbf{C}_i$ and $\mathbf{D}_i$ are state-space matrices with appropriate dimensions. The vector $\mathbf{w}(k) \in \mathbb{R}^n$ and $v(k) \in \mathbb{R}$ are assumed to be zero-mean normal distributed white noise satisfying

$$
E \left\{ \begin{bmatrix} \mathbf{w}(i) \\ v(i) \end{bmatrix} \begin{bmatrix} \mathbf{w}^T(j) & v^T(j) \end{bmatrix} \right\} = \begin{bmatrix} \mathbf{Q} & \mathbf{S} \\ \mathbf{S}^T & R \end{bmatrix} \delta_{ij} \geq 0
$$

which are statistically independent of the input vector $\mathbf{u}(k)$ and initial state $\mathbf{x}(0)$. It is assumed that the system matrices $\mathbf{A}_i$, $\mathbf{B}_i$, $\mathbf{C}_i$, $\mathbf{D}_i$, system order n, and matrices $\mathbf{Q}$, $\mathbf{S}$ and R are unknown *a priori*. The PS-based residual generator for each mode can be constructed using Eq. (2.18) as follows

$$
r_i(k) = \mathbf{v}_{s,i}(\mathbf{y}_s(k) - \mathbf{H}_{u,s,i}\mathbf{u}_s(k)),
\tag{4.5}
$$

where $\mathbf{u}_s(k)$ and $\mathbf{y}_s(k)$ are constructed in the same way as shown in Eq. (2.14). Equivalently, to avoid using the past data in FD scheme, the observer-based realization of FD system can be obtained using the one-to-one relationship between PS-based and observer-based residual generation. The residual signal $r_i(k)$ in Eq. (4.5) tends to be zero in fault-free cases, assuming that the system is operating in mode $\mathcal{M}_i$ and implicitly depends on the model parameters. Moreover, the residual signal $r_i(k)$ can be used to detect the faults which happen when system is working in mode $\mathcal{M}_i$ and for other modes it leads to false alarms.

The main idea is to identify the PS-based residual generator for each mode in the multimode system shown in Eq. (4.5) using EM algorithm and based on that, design the observer-based residual generator. Finally, fault detection will be achieved using a Bayesian inference strategy which combines the local fault detection results.

4.3 Identification of PS-based residual generator for multimode systems

Consider a dynamic system which is working in K different modes $\mathcal{M}_1, \mathcal{M}_2, \cdots, \mathcal{M}_K$, where each mode is characterized by Eq. (4.4) with different model parameters. The objective is to design a fault detection scheme for the proposed system using the PS-based residual generator in Eq. (4.5), assuming that the model parameters for different modes are unknown. For this purpose, first the dynamic mixture model should be identified using historical data. Assume that the available historical data $\mathcal{D}$ is collected from N different samples, each sample contains measurements of process variable $\mathbf{u} \in \mathbb{R}^l$ and quality variable $y \in \mathbb{R}$:

$$\begin{aligned}
\mathcal{D} &= \left\{ \begin{bmatrix} y(1) \\ \mathbf{u}(1) \end{bmatrix}, \begin{bmatrix} y(2) \\ \mathbf{u}(2) \end{bmatrix}, \cdots, \begin{bmatrix} y(N) \\ \mathbf{u}(N) \end{bmatrix} \right\} \\
&= \{ \mathbf{d}(1), \mathbf{d}(2), \cdots, \mathbf{d}(N) \}.
\end{aligned} \tag{4.6}$$

Under the assumption that the sample $\mathbf{d}(k)$ is generated in mode $\mathcal{M}_i$, i.e. $\mathbf{d}(k) \in \mathcal{M}_i$, the PS-based residual signal can be represented as

$$r_i(k) = \mathbf{v}_{s,i} \mathbf{y}_s(k) - \mathbf{v}_{s,i} \mathbf{H}_{u,s,i} \mathbf{u}_s(k) \sim \mathcal{N}(0, \sigma_i^2), \tag{4.7}$$

in fault-free case, where σ_i^2 represents the variance of the residual signal which depends on the noises in the system.

Therefore, the off-line design task of PS-based scheme is to estimate the following parameters using the historical data

$$\Theta = \{(\mathbf{v}_{s,1}, \mathbf{v}_{s,1}\mathbf{H}_{u,s,1}, \sigma_1^2), \cdots, (\mathbf{v}_{s,K}, \mathbf{v}_{s,K}\mathbf{H}_{u,s,K}, \sigma_K^2)\}, \qquad (4.8)$$

considering the elements of parity vectors as

$$\mathbf{v}_{s,i} = \begin{bmatrix} v_{s,i}^0 & v_{s,i}^1 & \cdots & v_{s,i}^s \end{bmatrix},$$

$$\mathbf{v}_{s,i}\mathbf{H}_{u,s,i} = \begin{bmatrix} \lambda_{s,i}^0 & \lambda_{s,i}^1 & \cdots & \lambda_{s,i}^s \end{bmatrix}. \qquad (4.9)$$

Using the maximum likelihood concept [11], the likelihood function can be expressed as

$$p(\mathbf{d}(k)) = p(y(k)|y(k-1), \cdots, y(k-s+1), \mathbf{u}(k), \cdots, \mathbf{u}(k-s+1)). \qquad (4.10)$$

In derivation of Eq. (4.10), it has been assumed that the current output of the dynamic system is described by the past outputs and temporal and past inputs up to s time delays. This is true since s is selected such that $s \geq n$ as mentioned in Section 2.2.3. Moreover it has been assumed that the rank condition in Eq. (2.15) holds true.

To extend it to multimode systems, the probability that a sample generated in a specific mode is integrated into the likelihood function through marginalization

$$p(\mathbf{d}(k)) = \sum_{i=1}^{K} p(\mathbf{d}(k)|\theta_i)p(\mathcal{M}_i), \qquad (4.11)$$

where $\theta_i = \{\mathbf{v}_{s,i}, \mathbf{v}_{s,i}\mathbf{H}_{u,s,i}, \sigma_i^2\}$. Using the MLE method, the unknown parameters Θ of the multimode system in Eq. (4.8) can be identified by maximizing following conditional likelihood function

$$\hat{\Theta}_{MLE} = \arg\max_{\Theta}\{p(\mathbf{d}(1), \cdots, \mathbf{d}(N)|\Theta)\}$$

$$= \arg\max_{\Theta}\{\prod_{k=1}^{N} p(y(k)|y(k-1),\cdots,y(k-s+1),$$

$$\mathbf{u}(k),\cdots,\mathbf{u}(k-s+1),\Theta)\}$$

$$= \arg\max_{\Theta}\{\prod_{k=1}^{N}\sum_{i=1}^{K} p(y(k)|y(k-1),\cdots,y(k-s+1),$$

$$\mathbf{u}(k),\cdots,\mathbf{u}(k-s+1),\theta_i)p(\mathcal{M}_i)\}.$$

$$(4.12)$$

The analytical solution of the MLE problem in Eq. (4.12) is not feasible. Therefore, in the same way that is proposed in Section 3.3.1, the EM algorithm is utilized to estimate the unknown parameters Θ.

The EM algorithm is based on the assumption that $\mathcal{D}$ is an incomplete data set in which the missing part can be interpreted as N tags, $\mathcal{Z} = \{z(1),\cdots,z(N)\}$, represents each sample generated in which operating mode. In EM algorithm, the conditional expectation of the log-likelihood for complete data $\mathcal{C} = \{\mathcal{D},\mathcal{Z}\}$ is calculated in E-step as follows

$$\mathcal{Q}(\Theta|\Theta^{old}) = E\{\log p(\mathcal{D},\mathcal{Z}|\Theta)|\mathcal{D},\Theta^{old}\} \qquad (4.13)$$

and the estimation of parameters in M-step is updated according to

$$\Theta = \arg\max_{\Theta} \mathcal{Q}(\Theta|\Theta^{old}) \qquad (4.14)$$

for ML.

The conditional expectation in the E-step of EM algorithm in Eq. (4.13) for the likelihood problem in Eq. (4.12), using the hidden variable $\mathcal{Z}$ as the mode indicator, can be written as

$$\mathcal{Q}(\Theta|\Theta^{old}) = E\{\log p(\mathcal{D},\mathcal{Z}|\Theta)|\mathcal{D},\Theta^{old}\}$$

$$= E\{\log p(\mathbf{d}(1),\cdots,\mathbf{d}(N),z(1),\cdots,z(N)|\Theta)|\mathcal{D},\Theta^{old}\}$$

$$= E\{\log \prod_{k=1}^{N} p(y(k)|y(k-1), \cdots, y(k-s+1), \mathbf{u}(k), \cdots,$$

$$\mathbf{u}(k-s+1), z(k), \Theta)p(z(k)|\Theta)|\mathcal{D}, \Theta^{old}\}$$

$$= E\{\sum_{k=1}^{N} \log p(y(k)|y(k-1), \cdots, y(k-s+1),$$

$$\mathbf{u}(k), \cdots, \mathbf{u}(k-s+1), z(k), \Theta)$$

$$+ \sum_{k=1}^{N} \log p(z(k)|\Theta)|\mathcal{D}, \Theta^{old}\}.$$

$$(4.15)$$

Taking the conditional expectation in Eq. (4.15) into consideration

$$\mathcal{Q}(\Theta|\Theta^{old}) = \sum_{k=1}^{N} \sum_{i=1}^{K} \{p(z(k) = i|\Theta^{old}, \mathcal{D}) \times$$

$$\log p(y(k)|y(k-1), \cdots, y(k-s+1), \mathbf{u}(k), \cdots,$$

$$\mathbf{u}(k-s+1), z(k), \theta_i)\}$$

$$+ \sum_{k=1}^{N} \sum_{i=1}^{K} p(z(k) = i|\Theta^{old}, \mathcal{D}) \times \log p(z(k) = i|\theta_i). \quad (4.16)$$

In the M-step of EM algorithm, the unknown parameters which maximize the conditional expectation $\mathcal{Q}(\Theta|\Theta^{old})$ are determined, given the parameters calculated in last iteration Θ^{old}. To this aim, the derivations of $\mathcal{Q}(\Theta|\Theta^{old})$ with respect to the unknown parameters are calculated and set to zero. The update of the parameters in the M-step is as follows

$$p(\mathcal{M}_i) = \frac{\sum\limits_{k=1}^{N} p(\mathcal{M}_i|\mathbf{d}(k), \theta_i^{old})}{N}$$

$$\sigma_i^2 = \frac{\sum\limits_{k=1}^{N} p(\mathcal{M}_i|\mathbf{d}(k), \theta_i^{old})(\mathbf{v}_{s,i}\mathbf{y}_s(k) - \mathbf{v}_{s,i}\mathbf{H}_{u,s,i}\mathbf{u}_s(k))^2}{\sum\limits_{k=1}^{N} p(\mathcal{M}_i|\mathbf{d}(k), \theta_i^{old})}$$

(4.17)

and by considering $v_{s,i}^s = 1$, the parity vectors are updated by solving following least squares problem

$$\begin{bmatrix} v_{s,i}^0 & v_{s,i}^1 & \cdots & v_{s,i}^{s-1} & \lambda_{s,i}^0 & \lambda_{s,i}^1 & \cdots & \lambda_{s,i}^s \end{bmatrix} \mathbf{G} = -\mathbf{H}, \quad (4.18)$$

where

$$\mathbf{G} = \left[\begin{array}{c|c} \mathbf{G}_1 & \mathbf{G}_2 \\ \hline \mathbf{G}_3 & \mathbf{G}_4 \end{array} \right], \quad \mathbf{H} = \left[\begin{array}{c|c} \mathbf{H}_1 & \mathbf{H}_2 \end{array} \right] \quad (4.19)$$

and the block elements of the matrices $\mathbf{G}_1 \in \mathbb{R}^{s \times s}$, $\mathbf{G}_2 \in \mathbb{R}^{s \times (s+1)l}$, $\mathbf{G}_3 \in \mathbb{R}^{(s+1)l \times s}$, $\mathbf{G}_4 \in \mathbb{R}^{(s+1)l \times (s+1)l}$, $\mathbf{H}_1 \in \mathbb{R}^s$ and $\mathbf{H}_2 \in \mathbb{R}^{(s+1)l}$ are calculated as follows

for $m, n = 1, \cdots, s$

$$\mathbf{G}_1(m, n) = \sum_{k=1}^{N} p(\mathcal{M}_i|\mathbf{d}(k), \theta_i^{old})y(k - n)y(k - m)$$

for $m = 1, \cdots, s$, and $n = 1, \cdots, s+1$

$$\mathbf{G}_2(m, n) = \sum_{k=1}^{N} p(\mathcal{M}_i|\mathbf{d}(k), \theta_i^{old})y(k - m)\mathbf{u}^T(k - n)$$

for $m = 1, \cdots, s+1$ and $n = 1, \cdots, s$

$$\mathbf{G}_3(m, n) = \sum_{k=1}^{N} p(\mathcal{M}_i|\mathbf{d}(k), \theta_i^{old})y(k - n)\mathbf{u}(k - m)$$

for $m, n = 1, \cdots, s + 1$

$$\mathbf{G}_4(m, n) = \sum_{k=1}^{N} p(\mathcal{M}_i | \mathbf{d}(k), \theta_i^{old}) \mathbf{u}(k - m) \mathbf{u}^T(k - n)$$

for $m = 1, \cdots, s$

$$\mathbf{H}_1(m) = \sum_{k=1}^{N} p(\mathcal{M}_i | \mathbf{d}(k), \theta_i^{old}) y(k - m) y(k)$$

for $m = 1, \cdots, s + 1$

$$\mathbf{H}_2(m) = \sum_{k=1}^{N} p(\mathcal{M}_i | \mathbf{d}(k), \theta_i^{old}) y(k) \mathbf{u}^T(k - m). \tag{4.20}$$

The probability $p(\mathcal{M}_i | \mathbf{d}(k), \theta_i^{old})$ in Eqs. (4.17) and (4.20) is calculated in the E-step using the Bayes' rule:

$$p(\mathcal{M}_i | \mathbf{d}(k), \theta_i^{old}) = \frac{p(\mathcal{M}_i) p(\mathbf{d}(k) | \mathcal{M}_i, \theta_i^{old})}{\displaystyle\sum_{j=1}^{K} p(\mathcal{M}_j) p(\mathbf{d}(k) | \mathcal{M}_j, \theta_j^{old})}. \tag{4.21}$$

Using the EM algorithm the parameters of the PS-based residual generator for different modes of the process is identified as shown in Eqs. (4.17), (4.18) and (4.21). Moreover the variance of the residual signals corresponding to each operating point is identified, which will be further used for threshold calculation. This residual signal can be used for fault detection purpose as will be explained in the forthcoming sections. Nevertheless, the on-line realization of the PS-based residual generator entails incorporation of past and temporal data, which might not be appealing from implementation point of view.

To overcome this shortcoming, the one-to-one relationship between PS-based residual generator and DO, introduced in Section 2.2.4 is used to construct a diagnostic observer for each mode. Following the form of DO in Eq. (2.11), the multimode observer-based residual generation is constructed in the following form

$$\mathbf{z}(k+1) = \mathbf{A}_z\mathbf{z}(k) + \mathbf{B}_{z,i}\mathbf{u}(k) + \mathbf{L}_{z,i}y(k)$$
$$r_i(k) = g_{z,i}y(k) - \mathbf{c}_z\mathbf{z}(k) - \mathbf{d}_{z,i}\mathbf{u}(k), \qquad (4.22)$$

where

$$\mathbf{A}_z = \begin{bmatrix} 0 & 0 & \cdots & 0 & 0 \\ 1 & 0 & \cdots & 0 & 0 \\ \vdots & \vdots & \ddots & \vdots & \vdots \\ 0 & 0 & \cdots & 1 & 0 \end{bmatrix}, \quad \mathbf{L}_{z,i} = -\begin{bmatrix} v_{s,i}^0 \\ v_{s,i}^1 \\ \vdots \\ v_{s,i}^{s-1} \end{bmatrix}$$

$$\mathbf{c}_z = \begin{bmatrix} 0 & \cdots & 0 & 1 \end{bmatrix}, \quad g_{z,i} = v_{s,i}^s$$

$$\mathbf{d}_{z,i} = \lambda_{s,i}^s, \ \mathbf{B}_{z,i} = \begin{bmatrix} \lambda_{s,i}^0 \\ & \lambda_{s,i}^1 \\ & & \ddots \\ & & & \lambda_{s,i}^{s-1} \end{bmatrix}. \qquad (4.23)$$

The residual signals obtained by the DO or PS-based residual generators indicate the faults under the assumption that the collected measurements are generated in the corresponding mode. To construct an indicator which represents the fault in the system and distinguishes it with mode changes, the Bayesian inference strategy is used to combine these hypotheses and define a global index for FD purpose.

4.4 Fault Detection Scheme

So far the estimation of the finite mixture model is described. For fault detection purpose, the parity vectors corresponding to each component is identified and based on them the observer based residual generator can be constructed using Eqs. (4.22) and (4.23). This step is done in off-line design of fault detection system.

For on-line implementation, when a new sample of measurement is available, the fault detection system should firstly assign the sample to a component, which means the sample is generated from which model or the system is working in which mode, and then based on it, decide if the sample is faulty or normal assuming that it originally came from that mode. A Bayesian inference scheme can be utilized in this step to perform this analysis. First the posterior probability that a sample belongs to each component will be calculated

$$p(\mathbf{d}(k) \in \mathcal{M}_i) = p(\mathcal{M}_i | \mathbf{d}(k)) = \frac{p(\mathbf{d}(k) | \mathcal{M}_i) p(\mathcal{M}_i)}{p(\mathbf{d}(k))}$$

$$= \frac{p(\mathbf{d}(k) | \mathcal{M}_i) p(\mathcal{M}_i)}{\sum\limits_{i=1}^{K} p(\mathcal{M}_i) p(\mathbf{d}(k) | \mathcal{M}_i)}, \tag{4.24}$$

where $\sum\limits_{i=1}^{K} p(\mathcal{M}_i | \mathbf{d}(k)) = 1$ and $\mathbf{d}(k) = \begin{bmatrix} y(k) \\ \mathbf{u}(k) \end{bmatrix}$.

Assuming that $\mathbf{d}(k) \in \mathcal{M}_i$ the residual corresponding to this sample can be calculated using Eq. (4.22) where the residual signal is a Gaussian signal and $r_i \sim \mathcal{N}(0, \sigma_i^2)$ in normal operating conditions. Based on this residual, the following test signal

$$J_i(k) = \frac{1}{\sigma_i^2} (r_i(k))^T r_i(k) \tag{4.25}$$

can be constructed for fault detection purposes which follows χ^2 distribution with one degree of freedom and serves as an indicator for faults assuming that the measurement signal comes from component $\mathcal{M}_i$. For the purpose of fault detection, a global index is further defined to represent the probability that the sample is subjected to a fault:

$$J_g(k) = p(\mathbf{d}(k) \in f). \tag{4.26}$$

This index can be obtained through marginalization as

$$J_g(k) = \sum_{i=1}^{K} p(\mathbf{d}(k) \in f | \mathbf{d}(k) \in \mathcal{M}_i) p(\mathbf{d}(k) \in \mathcal{M}_i), \tag{4.27}$$

where

$$p(\mathbf{d}(k) \in f | \mathbf{d}(k) \in \mathcal{M}_i) = p(J_i \leq J_i(k)). \qquad (4.28)$$

The probability in (4.28) can be obtained by integrating the χ^2 probability density function with one degree of freedom. Since $0 \leq J_g(k) \leq 1$, a confidence level $(1 - \alpha)$ can be specified for fault detection purpose with the hypothesis as follows.

$$\begin{cases} J_g \leq 1 - \alpha & \text{fault free} \\ J_g > 1 - \alpha & \text{faulty} \end{cases} \qquad (4.29)$$

The procedure for design and implementation of the proposed fault detection scheme can be summarized in Algorithm 3.

Algorithm 3. *Design of fault detection system*

Step 1 *Collect N samples of data vector as shown in Eq. (4.6).*

Step 2 *Perform EM algorithm to estimate the mixture model parameters from Eqs. (4.16), (4.17) and (4.20).*

Step 3 *For $i = 1 : K$ construct the diagnostic observer corresponding to each component using Eq. (4.23).*

Step 4 *When a new sample of measurements is available, for $i = 1 : K$ calculate the probability shown in (4.24).*

Step 5 *For $i = 1 : K$ compute the residual associated to each component as shown in Eqs. (4.22) and (4.25).*

Step 6 *Calculate the global indicator which represents the probability of the sample being faulty as shown in Eq. (4.27).*

Step 7 *Use the fault detection hypothesis in (4.29) to detect fault and go to step 4.*

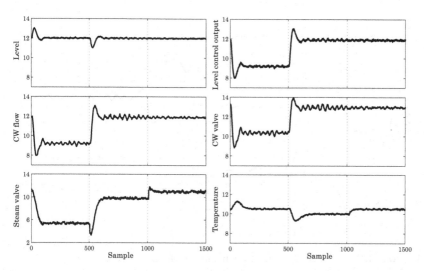

Figure 4.1: Process measurements used for off-line training step

4.5 An illustrative example

In this section, the proposed method for fault detection in nonlinear dynamic multimode systems is applied to the CSTH benchmark example. The CSTH process considered in this demonstrative study operates in three different operating regimes as shown in Table 3.1. For each mode 500 samples are acquired to carry out this experiment. The process measurements which are used for off-line training in this study are shown in Fig. 4.1. The PS-based residual generator is identified as described in Eqs. (4.15) to (4.21). Furthermore, using the one-to-one relationship between PS-based residual generator and DO, the observer-based residual generator is designed for all K different modes, as shown in Eqs. (4.22) and (4.23).

In on-line monitoring step of this example, one fault scenario is considered, namely fault in hot water valve f_2. The on-line process measurements are shown in Fig. 4.2. For each mode 200 samples are used. The transient behavior of the system due to set-point changes

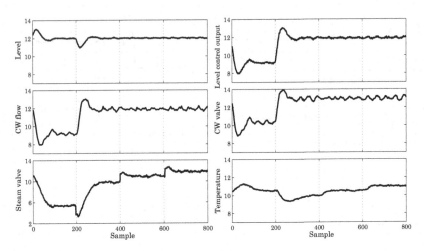

Figure 4.2: On-line process measurements for fault detection

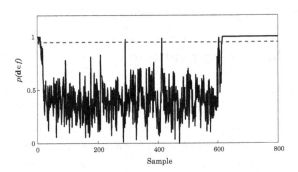

Figure 4.3: Fault detection indicator

is also demonstrated in the measurement plots. The fault f_2 appears after $600th$ sample and continues till end of this simulation.

Further, the observer-based residual signals are combined with the probability that the measured sample belongs to the corresponding mode to build up the global fault detection indicator as described in Eq. (4.27). The result is depicted in Fig. 4.3. The 95% confidence level is represented by the red-dashed line. At the beginning of the simulation the FD index is over the threshold and indicating a false alarm, which is due to the error dynamics in the residual generator. The FD index remains below the threshold till $600th$ sample where the fault is happening (except for few false alarms). After appearance of the fault in the process, the residual cross the threshold and remains over the threshold, which indicates the successful detection of the fault.

4.6 Concluding remarks

In this chapter, the fault detection scheme for nonlinear multimode systems, proposed in Chapter 3 is extended to dynamic systems. An identification method based on EM algorithm is derived which directly identifies the parity vectors associated with different operating modes. Then, the identified parity vectors are used to design a multi-observer scheme for residual generation. In on-line monitoring, when a new sample of measurements is available, the residual signals corresponding to each mode is combined with the hypothesis that the sample is generated in that mode, to construct a global index for FD purpose. The performance and effectiveness of the method are demonstrated through the simulation on CSTH SIMULINK benchmark.

5 Fault diagnosis in multimode nonlinear processes

Fault isolation plays a central role in process monitoring and diagnosis and sometimes is a real challenge for process engineers. Basically, fault isolation is the task of gaining process information about the location of the fault in the process [29]. In a modern large scale process where the number of faults, process components and measurements are huge the fault isolation becomes very complicated.

In the last decades, several solutions for fault isolation using MSPM techniques have been proposed. Fisher discriminant analysis has been used for fault diagnosis [20, 21, 47, 59, 131] and applied in various range of applications. Isolation enhanced PCA has been developed which delivers structured residuals where each is sensitive to some preselected sensor or actuator fault [43, 41, 42]. Contribution analysis has been widely used to determine the contributions of a variable to a fault detection index [83, 87, 117]. The variables with high contributions are responsible for the undesired behavior of the system. The contribution analysis methods may not explicitly determine the cause of fault alarm, but it can be used as a guideline for process engineers to find the source of the fault.

Fault isolation methods based on MSPM techniques usually rely on uni-modal Gaussian distribution of normal operating data. Recently, few research studies have addressed fault isolation issues in multimode processes. In [16], the authors have developed a probabilistic contribution analysis method based on missing variable approach. Once a fault is detected, the monitoring index will be recalculated with one variable being missing. This will be repeated for all variables. The variable corresponding to the smallest recalculated index will be denoted as the risky variable. The proposed idea has been extended to PPCA mixture model [70, 105] for fault detection and diagnosis in multimode processes.

Motivated by the above mentioned works, a new probabilistic approach for fault diagnosis in multimode processes is developed and presented in this chapter. For that, the unified index proposed in Chapter 3 is used for quality related fault detection purpose. Once a fault is detected, the index will be decomposed into two parts, one represents the behavior of monitoring index in normal operating conditions and the other represents the contribution of the variables to the fault which will be used for diagnosis purpose.

5.1 Preliminaries

In practice, fault detection is usually followed by the isolation step where location of the fault is determined. In context of MSPM, isolation is usually accomplished with contribution analysis where the process variables contributing to the fault are determined and a contribution plot is constructed. The statistical model which is built by MSPM methods will be used for analysis of the contribution of process variables or latent variables on the fault detection indices. Methods based on complete and partial decomposition and angle based contribution analysis have been developed and successfully applied for diagnosis purposes. In these approaches usually the quadratic form of fault detection index is considered:

$$\text{Index}(\mathbf{x}) = \bar{\mathbf{x}}^T \mathbf{D} \bar{\mathbf{x}}, \qquad (5.1)$$

where $\bar{\mathbf{x}} \in \mathbb{R}^m$ is the normalized process measurement and the matrix $\mathbf{D}$ is constructed based on the monitoring index and the MSPM method, for instance in PCA-based process monitoring

$$\mathbf{D} = \mathbf{P}_{pc} \mathbf{\Lambda}_{pc}^{-1} \mathbf{P}_{pc}^T \qquad (5.2)$$

for T^2 (see Eq. (2.25)) and

$$\mathbf{D} = \mathbf{P}_{res} \mathbf{\Xi} \mathbf{P}_{res}^T \qquad (5.3)$$

for T^2_{new} (see Eq. (2.29)). The monitoring index can be decomposed as

$$\text{Index}(\mathbf{x}) = \bar{\mathbf{x}}^T \mathbf{D} \bar{\mathbf{x}} = ||\mathbf{D}^{(1/2)} \bar{\mathbf{x}}||^2 = \sum_{j=1}^{m} \left(\boldsymbol{\xi}_j^T \mathbf{D}^{(1/2)} \bar{\mathbf{x}} \right)^2 = \sum_{j=1}^{m} c_j^{\text{Index}},$$
(5.4)

where c_j^{Index} is the contribution of the variable x_j to Index($\mathbf{x}$) and $\boldsymbol{\xi}_j$ is the jth column of identity matrix [83]. By choosing appropriate $\mathbf{D}$ the diagnosis scheme for PCA and PLS method can be achieved [77].

Recently, in [1] it has been revealed that the standard contribution analysis may lead to misdiagnosis of the faults and an alternative method has been proposed. The method is based on the reconstruction of a fault detection index along a variable direction, hence it is called reconstruction based contribution (RBC). When a fault happens in the system with direction $\boldsymbol{\xi}_j$, the reconstructed measurement vector can be represented as:

$$\mathbf{z}_j = \bar{\mathbf{x}} - \boldsymbol{\xi}_j \mathbf{f}, \quad (5.5)$$

where $\mathbf{f}$ is the reconstructed part to be determined and $\mathbf{z}_j$ represents the fault-free behavior of variables and can be constructed by finding the value of $\mathbf{f}$ which minimizes Index($\mathbf{z}_j$)

$$\text{Index}(\mathbf{z}_j) = \mathbf{z}_j^T \mathbf{D} \mathbf{z}_j = ||\mathbf{D}^{(1/2)}(\bar{\mathbf{x}} - \boldsymbol{\xi}_j \mathbf{f})||^2. \quad (5.6)$$

The optimal value for $\mathbf{f}$ can be obtained by derivation of Index($\mathbf{z}_j$) with respect to $\mathbf{f}$:

$$\frac{d(\text{Index}(\mathbf{z}_j))}{d\mathbf{f}} = -2(\bar{\mathbf{x}} - \boldsymbol{\xi}_j \mathbf{f})^T \mathbf{D} \boldsymbol{\xi}_j. \quad (5.7)$$

Equating the Eq. (5.7) to zero yields to:

$$\mathbf{f} = (\boldsymbol{\xi}_j^T \mathbf{D} \boldsymbol{\xi}_j)^{-1} \boldsymbol{\xi}_j^T \mathbf{D} \bar{\mathbf{x}}. \quad (5.8)$$

The reconstruction-based contribution of the variable x_j to fault detection index Index($\mathbf{x}$) can be described by

$$RBC_j^{\text{Index}} = ||\mathbf{D}^{(1/2)}\boldsymbol{\xi}_j\mathbf{f}||^2 \qquad (5.9)$$

or

$$RBC_j^{\text{Index}} = \bar{\mathbf{x}}^T\mathbf{D}\boldsymbol{\xi}_j(\boldsymbol{\xi}_j^T\mathbf{D}\boldsymbol{\xi}_j)^{-1}\boldsymbol{\xi}_j^T\mathbf{D}\bar{\mathbf{x}}. \qquad (5.10)$$

It is interesting to point out that following relation exists between fault detection index, reconstructed index and reconstruction based contribution:

$$\text{Index}(\mathbf{x}) = \text{Index}(\mathbf{z}_j) + RBC_j^{\text{Index}}. \qquad (5.11)$$

Both reconstructed index, Index($\mathbf{z}_j$) and RBC_j^{Index} can be used for diagnosis purpose [1, 34].

Based on contribution analysis concept, several methods have been developed and applied for different applications. An overview of these methods and their generalization is given in [2] and references therein.

5.2 Probabilistic fault diagnosis in multimode processes

Contribution analysis methods usually assume a single normal operating mode for the plant. In this context, an abnormal event will form a new operating region and the difference between normal and faulty states is used to identify the variable contribution. In many real applications the process itself works in different operating region and using standard contribution analysis methods will lead to misdiagnosis of the faults. To solve this problem, in this section a new fault isolation method is proposed which follows the FD methods proposed in Chapters 3 and 4 and tries to represent the variable contributions in a probabilistic form.

To extend the fault isolation approaches to multimode cases, the fault detection indices are generalized as an index which represents the probability of fault $p(\mathbf{x}(k) \in f)$ given a sample of measurement $\mathbf{x}(k)$. Using marginalization, the above mentioned probability can be represented as:

$$p(\mathbf{x}(k) \in f) = \sum_{i=1}^{K} p(\mathbf{x}(k) \in f | \mathbf{x} \in \mathcal{M}_i) p(\mathbf{x} \in \mathcal{M}_i), \qquad (5.12)$$

where $p(\mathbf{x}(k) \in f | \mathbf{x}(k) \in \mathcal{M}_i)$ can be calculated by integrating the probability density function of the detection index up to its current value. In other words:

$$p(\mathbf{x}(k) \in f | \mathbf{x}(k) \in \mathcal{M}_i) = p(\text{Index}(\mathbf{x}, i) \leq \text{Index}(\mathbf{x}(k), i))$$
$$= \int_{0}^{\text{Index}(\mathbf{x}(k), i)} \text{pdf}(\text{Index}(\mathbf{x}, i)) \mathrm{d}\mathbf{x}, \quad (5.13)$$

where $\text{Index}(., i)$ represents the calculated value of index based on the assumption that the sample belongs to $\mathcal{M}_i$. Equation (5.13) also serves as a local fault indicator for mode $\mathcal{M}_i$. Since $0 \leq p(\mathbf{x}(k) \in f) \leq 1$, a confidence level $(1 - \alpha)$ can be specified for fault detection purpose with the hypothesis as follows.

$$\begin{cases} p(\mathbf{x}(k) \in f) \leq 1 - \alpha & \text{fault free} \\ p(\mathbf{x}(k) \in f) > 1 - \alpha & \text{faulty} \end{cases} \qquad (5.14)$$

The main idea of this new fault isolation approach is to calculate the contribution of a faulty measurement sample $\mathbf{x}(k)$, assuming that it belongs to mode $\mathcal{M}_i$ and then combine it with the hypothesis that the measurement $\mathbf{x}(k)$ is generated under the mode $\mathcal{M}_i$. The local fault detection index, $\text{Index}(\mathbf{x}(k), i)$ can be decomposed using Eq. (5.11) as

$$\text{Index}(\mathbf{x}(k), i) = \text{Index}(\mathbf{z}_j(k), i) + RBC_{j,i}^{\text{Index}}, \qquad (5.15)$$

where $\text{Index}(\mathbf{z}_j(k), i)$ is the detection index according to the reconstructed measurement along variable $x_j(k)$, assuming $\mathcal{M}_i$ as the current operating mode and $RBC_{j,i}^{\text{Index}}$ is the amount of reconstruction based contribution for that. Using Eq. (5.15), the Eq. (5.13) can be rewritten as

$$p(\mathbf{x}(k) \in f | \mathbf{x}(k) \in \mathcal{M}_i) = \int_0^{\text{Index}(\mathbf{x}(k),i) - \text{Index}(\mathbf{z}_j(k),i)} \text{pdf}(\text{Index}(\mathbf{x}, i))\mathrm{d}\mathbf{x}$$
$$+ \int_{\text{Index}(\mathbf{x}(k),i) - \text{Index}(\mathbf{z}_j(k),i)}^{\text{Index}(\mathbf{x}(k),i)} \text{pdf}(\text{Index}(\mathbf{x}, i))\mathrm{d}\mathbf{x}.$$

(5.16)

The first term in right hand side of (5.16) represents the effects reconstructed contribution into the local fault probability and the second term represents the contribution of reconstructed variable to the local fault probability. Moreover the first term in Eq. (5.16) can be expressed as

$$PRBC_{j,i}^{\text{Index}} = \int_0^{\text{Index}(\mathbf{x}(k),i) - \text{Index}(\mathbf{z}_j(k),i)} \text{pdf}(\text{Index}(\mathbf{x}, i))\mathrm{d}\mathbf{x}$$
$$= p(0 \le \text{Index}(\mathbf{x}, i) \le \text{Index}(\mathbf{x}(k), i) - \text{Index}(\mathbf{z}_j(k), i) | \mathbf{x}(k) \in \mathcal{M}_i),$$

(5.17)

where $PRBC$ stands for probabilistic RBC.

Generalizing it to the multimode processes, a $PRBC$ can be defined as:

$$PRBC_j^{\text{Index}} = \sum_{i=1}^{K} PRBC_{j,i}^{\text{Index}} p(\mathbf{x}(k) \in \mathcal{M}_i)$$
$$= \sum_{i=1}^{K} p(0 \le \text{Index}(\mathbf{x}, i) \le \text{Index}(\mathbf{x}(k), i)$$
$$- \text{Index}(\mathbf{z}_j(k), i) | \mathbf{x}(k) \in \mathcal{M}_i) p(\mathbf{x}(k) \in \mathcal{M}_i).$$

(5.18)

The posterior probability $p(\mathbf{x}(k) \in \mathcal{M}_i)$ is used in Eq. (5.18) to incorporate the contribution of each local model to the $PRBC_j^{\text{Index}}$. It is worth pointing out that $0 \le PRBC_j^{\text{Index}} \le 1$ and the variable

$x_j(k)$ with the highest $PRBC_j^{\text{Index}}$ has more contribution to the fault and possibly represents the source of the malfunction in the system.

To calculate the probability $p(\mathbf{x}(k) \in \mathcal{M}_i)$ the posterior probability that the reconstructed measurement $\mathbf{z}_j(k)$ belongs to the mode $\mathcal{M}_i$ is used. This is due to the fact that $\mathbf{x}(k)$ is assumed to be faulty measurement, therefore may not provide a correct representation of the actual operating mode of the process. Therefore the reconstructed measurement $\mathbf{z}_j(k)$ which represents the estimation of fault free measurement assuming that the fault happens in the jth sensor, is replaced in marginalization. In the same way as shown in Section 3.3.2, the Bayesian inference strategy is used to calculate this posterior probability

$$p(\mathbf{z}_j(k) \in \mathcal{M}_i) = \frac{p(\mathbf{z}_j(k)|\mathcal{M}_i)p(\mathcal{M}_i)}{p(\mathbf{z}_j(k))} = \frac{w_i g(\mathbf{z}_j(k)|\theta_i)}{\sum\limits_{l=1}^{K} w_l g(\mathbf{z}_j(k)|\theta_l)}. \quad (5.19)$$

Once a fault is detected using the FD-index in (5.12), the reconstructed index, $\text{Index}(\mathbf{z}_j(k), i)$, will be calculated using (5.11) and inserted in (5.18) together with the $\text{Index}(\mathbf{x}(k), i)$ and the posterior probability $p(\mathbf{z}_j(k) \in \mathcal{M}_i)$, to calculate the probabilistic contribution of variables to the fault, and the variables corresponding to the largest contributions are the risky variables.

5.3 An illustrative example

To study the effectiveness and evaluate the performance of the proposed probabilistic contribution analysis, it is applied on the CSTH benchmark. The simulation here follows the fault detection step shown in Section 3.4. The fault has been successfully detected in samples 150 to 200 and the results has been shown in Fig. 3.3.

The task of contribution analysis is to determine the contribution of each variable into the fault detection index $p(\mathbf{x}(k) \in f)$ given the current measurements $\mathbf{x}(k)$. Assuming that the plant is working on mode $\mathcal{M}_i$, $\text{Index}(\mathbf{x}(k), i)$ is calculated using Eq. (3.4) and the

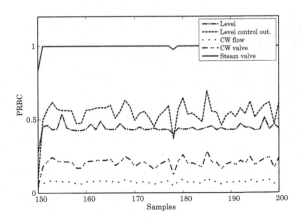

Figure 5.1: $PRBC_j^{\text{Index}}$ for fault f_2 in CSTH benchmark

model parameters for $\mathcal{M}_i$. Index($\mathbf{z}_j(k), i$) are calculated for $j = 1, \cdots, m$ using Eqs. (5.5) and (5.6). These indices are inserted in Eq. (5.17) to calculate $PRBC_{j,i}^{\text{Index}}$ and further integrated with the mode probabilities obtained by Eq. (5.19) to form the $PRBC_j^{\text{Index}}$ shown in Eq. (5.18).

The calculated values for $PRBC_j^{\text{Index}}$ for the fault scenario f_2 according to Section 3.4 is depicted in Fig. 5.1 for each variable $x_j(k)$ for $k = 150, \cdots, 200$ where the fault is detected in the system.

To make the representation similar to the classical contribution plot, the bar plot of the mean values of the $PRBC_j^{\text{Index}}$ is shown in Fig. 5.2. It can be seen from Figs. 5.1 and 5.2 that the steam valve measurement signal has the highest contribution to the fault detection index and therefore the most probable source of the deviation in the temperature signal. Indeed, the source or f_2 was an actuator fault in steam valve which approves the simulation results.

5.4 Concluding remarks

This chapter provides a novel solution for fault diagnosis and isolation in nonlinear multimode systems. This is achieved by studying the

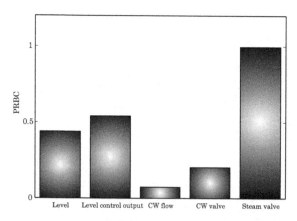

Figure 5.2: Bar plot of $PRBC_j^{\text{Index}}$ for fault f_2 in CSTH benchmark

reconstruction based contribution analysis and extending the concept to the multimode process. To this end, the process variables are reconstructed in different modes to form the local contributions. The local contributions are combined with the hypothesis that the reconstructed variable belongs to a specific mode and is marginalized to form a global contribution for each variable.

Although the computational complexity of the method is higher than standard contribution analysis, due to calculation of contributions in different modes, its effectiveness in multimode process applications is dominant compared to traditional methods.

6 Bayesian approach for fault treatment

After the successful fault diagnosis, the faulty system should be recovered. A technical system is made up of many different interconnected components, where the symptoms of a specific problem may have different causes. Furthermore, a failure in a component may result in several consequences. Thus, detection of the root causes of alarms and applying the correct maintenance operation are always crucial tasks in a complex industrial system. Therefore, it is necessary to have a decision support system which helps the plant engineer to discover the causes of the alarms through the troubleshooting process and provide a systematic search for the root cause of a problem in a process to make the process operational again.

In this framework, the traditional way in automatic control is to employ the fault tolerant systems to increase the availability of the system in presence of faults. This is usually achieved by means of control reconfiguration to increase the dependability of the system, known as fault tolerant control (FTC) [9]. However, it cannot always provide the desired performance and sometimes induces more losses and a fault tolerant system cannot be achieved for every fault.

To this aim, in this chapter a new probabilistic approach is introduced to design a decision support system. The main idea in this approach is to combine the probability of all possible faults in the system with the economical aspect of each possible corrective operation. This provides a list of corrective operations ranked based on their costs and their effects on the overall performance of the system [46].

6.1 Preliminaries and problem formulation

A large-scale technical system consists of several thousand components and hundreds of control loops. When a malfunction in the system or a degradation in product quality is detected, the operator has to find the cause of the fault by analysing the process measurements and fault detection monitors. The time required to isolate and treat the fault results in unplanned interruption of the production and is the main cause of losses in the production systems. An automated decision support system can diminish the losses by helping the operator to identify the faulty component and reducing the unplanned shutdown time.

In many industrial applications, the process measurements are recorded in process historian and are well documented. The time intervals where the process is subjected to different faults and root cause of the faults are described. This information can be used to analyze the impacts of the faults and in the case where the same fault happens in future, be utilized for corrective action generation. To this end, the decision support system should be able to learn the faulty models from available historical data and at each time, estimates which faults are happening in the system. For that, a method proposed here which formulates this prediction as probability that a certain fault is happening using available information. Using the process historical data, one can build the statistical models from faulty and fault-free measurements and use it for estimation of fault probability. Together with economical restriction of the corrective operations and their influence on the overall performance of the system, a decision support system can be designed, which can help the process engineer to decide about the most proper maintenance operation.

6.2 Estimation of the fault probability

Consider the following example shown in Fig. 6.1, where the system is subjected to two different fault scenarios f_1 and f_2. The faults are affecting the process measurements $\mathbf{x}$. Moreover, they affect other

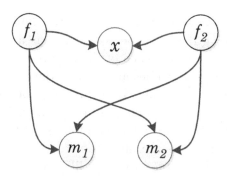

Figure 6.1: An example of Bayesian network

monitoring indices in the system m_1, m_2 which are the results of FDD system.

The task is to use Bayesian method to synthesize the monitoring indices and the data to design an optimal decision making system [52]. For this purpose, the probability of the faults f_1 and f_2 using the on-line values for $\mathbf{x}$, m_1, m_2 are calculated, following the idea proposed in [52]. In other words, following conditional probabilities should be taken into account:

$$p(f_1|\mathbf{x}(k), m_1(k), m_2(k))$$
$$p(f_2|\mathbf{x}(k), m_1(k), m_2(k)) \tag{6.1}$$

The scheme consists of two steps: off-line modeling and on-line implementation. In off-line modeling, the statistical models for different faulty episodes $p(\mathbf{x}|f_1, f_2)$ and $p(m_1, m_2|f_1, f_2)$, and corresponding *a priori* probability $p(f_1, f_2)$ are obtained from available historical data. In on-line implementation, the probabilities in Eq. (6.1) are calculated using the current process measurements $\mathbf{x}(k)$ and monitor readings $m_1(k)$, $m_2(k)$ together with *a priori* probability $p(f_1, f_2)$. The *a priori* probability of the faults can be determined from historical data or the performance of the components. An example for the possible results of off-line training step is shown in Table 6.1, considering that the data follow Gaussian distribution for each episode.

Assuming the Gaussian distribution for the data for each episode, the off-line training data can be considered as a finite mixture of different components where each of them follows Gaussian distribution with different mean values and/or covariance matrices. Several methods can be used to learn such mixture models which can be mainly categorized in supervised and unsupervised learning methods where in the former, the associated data for each episode is known and in the later, the data sets are unclassified [36, 124]. For supervised learning, methods based on maximum likelihood estimation (MLE) are popular and for unsupervised learning expectation-maximization (EM) method can be used [82]. MLE tries to estimate the unknown parameters $\Theta = \{m_i, \Sigma_i\}$, the mean and covariance of Gaussian component i, by determining the parameter which maximize following log-likelihood function:

$$\hat{\Theta}_{MLE} = \arg\max_{\Theta}\{\log p(\mathbf{X}|\Theta)\}, \qquad (6.2)$$

and in EM algorithm, the conditional expectation of the log-likelihood for complete data $C = \{X, Z\}$ is calculated in E-step as follows [81]

$$\mathcal{Q}(\Theta|\Theta^{old}) = E\{\log p(\mathbf{X}, \mathcal{Z}|\Theta)|\mathbf{X}, \Theta^{old}\} \qquad (6.3)$$

and updates the estimation of parameters in M-step according to

$$\hat{\Theta}_{EM} = \arg\max_{\Theta} \mathcal{Q}(\Theta|\Theta^{old}) \qquad (6.4)$$

For more information about EM algorithm and its derivation the readers are referred to Appendix A. The estimations $\hat{\Theta}_{MLE}$ or $\hat{\Theta}_{EM}$ can be used to calculate $p(\mathbf{x}|f_1, f_2)$ in Table 6.1.

The main advantage of this scheme is its ability to deal with uncertainties in the process and for the cases where a certain fault has different symptoms and affects different monitoring indices [52, 89]. The inherent property of Bayesian methods to incorporate the expert knowledge can help the decision making system to use the operator knowledge about a certain event in the system. Moreover, for the cases where readings for some monitoring indices are not available, this scheme can still provide correct results with high confidence which is a great significance in practice [90].

6.3 Decision support system

To consider the uncertainties in FDD and prognosis in corrective operation generation and decision making, the probabilities obtained previously can be used. The probability of occurrence of a specific fault or degradation in overall performance of the system, together with loss minimization technique which reflects maintenance operations, form the basis for this decision support scheme. When the faulty measurements are also available in historical data, the faulty model can be extracted from that and used in on-line implementation phase to calculate the probability of a certain fault which reflects abnormalities in component or subsystem level as described previously. Associated with each fault there will be a list of corrective operations ranging from doing nothing to the component replacement or repair and controller tuning, with their costs and benefits on improvement of system performance. Cost minimization technique has been the traditional way to determine the maintenance action [13, 26, 97]. The main idea here is to combine the probability that a certain fault is happening and the losses corresponding to the corrective operations respective to that fault to find the most proper corrective operation(s) among the list of all operations.

A loss due to a corrective operation is defined here as [12]

$$\text{Loss}(CA_j, f_i) = \text{Cost}(CA_j) + \text{Loss}(f_i) - \text{Benefit}(CA_j, f_i) \quad (6.5)$$

where CA_j represents the corrective operation j associated with the fault i, f_i. Since $0 \leq \text{Benefit}(CA_j, f_i) \leq \text{Loss}(f_i)$, Eq. (6.5) can be written as

$$\text{Loss}(CA_j, f_i) = \text{Cost}(CA_j) + (1 - \alpha_{i,j})\text{Loss}(f_i)$$
$$= \text{Cost}(CA_j) + \text{Loss}(CA_j|f_i) \quad (6.6)$$

where $0 \leq \alpha_{i,j} \leq 1$ is the normalized value of the benefit of CA_j and can be interpreted as the following conditional probability

$$\alpha_{i,j} = p(\text{Benefit}(CA_j, f_i) = \text{high}|f_i, CA_j) \quad (6.7)$$

The parameter $\alpha_{i,j}$ will be determined by expert knowledge. Moreover, it might be possible to obtain $\alpha_{i,j}$ from historical data. It is worth pointing out that, $\text{Cost}(CA_j)$ in Eq. (6.6) represents the fixed costs of the CA_j which does not depend on the faults, whilst $\text{Loss}(CA_j|f_i)$ depends on fault and CA_j.

Following risk function is defined for each corrective operation

$$R(CA_j) = \text{Cost}(CA_j) + \sum_{f_i} \text{Loss}(CA_j|f_i)p(f_i). \qquad (6.8)$$

The CA_j corresponding to minimum risk represents the most proper operation.

6.4 Probabilistic decision support system

Another approach to determine the proper maintenance operation is to define the optimization in a probability form and use the maximum a *posteriori* probability (MAP) criterion to find CA_j which has the highest a *posteriori* probability. For that the best corrective operation can be defined as the CA_j which has the highest impact on system performance and the lowest loss. In other words, the most proper corrective operation is the one associated with highest probability stated below:

$$\hat{CA}_{MAP} = \arg\max_{j}\{p(\text{Benefit}(CA_j, f_i) = \text{high},$$

$$\text{Loss}(CA_j, f_i) = \text{low}|CA_j)\} \qquad (6.9)$$

The probability in Eq. (6.9) can be written as

$$p(\text{Benefit}(CA_j, f_i) = \text{high}, \text{Loss}(CA_j, f_i) = \text{low}|CA_j)$$

$$= \sum_{f_i} p(\text{Benefit}(CA_j, f_i) = \text{high}|f_i, CA_j) \times$$

$$p(\text{Loss}(CA_j, f_i) = \text{low}|f_i, CA_j) \times p(f_i) \qquad (6.10)$$

The first term on the right hand side of Eq. (6.10) is $\alpha_{i,j}$ introduced in Eq. (6.7) which can be obtained either by expert knowledge or from

Table 6.1: The conditional probability table and distributions for CSTH example

Fault			Evidences						$p(\mathbf{x}\|f_1, f_2)$
			$p(m_1\|f_1, f_2)$			$p(m_2\|f_1, f_2)$			
f_1	f_2	$p(f)$	L	M	H	L	M	H	
0	0	77%	98.2%	1.6%	0.2%	92.6%	6.8%	0.6%	$\mathbf{x} \sim \mathcal{N}(m_N, \Sigma_N)$
0	1	12%	2.3%	91.4%	6.3%	6.6%	12.1%	81.3%	$\mathbf{x} \sim \mathcal{N}(m_{f_1}, \Sigma_{f_1})$
1	0	11%	89.3%	7.2%	3.5%	0.7%	0.1%	99.2%	$\mathbf{x} \sim \mathcal{N}(m_{f_2}, \Sigma_{f_2})$

historical data. The probability $p(\text{Loss}(CA_j, f_i) = \text{low}|f_i, CA_j)$ can be formulated as $1 - p(\text{Loss}(CA_j, f_i) = \text{high}|f_i, CA_j)$, where

$$\text{Loss}(CA_j, f_i) = \text{high}|f_i, CA_j$$
$$\sim \mathcal{U}\left(\text{Cost}(CA_j), \text{Cost}(CA_j) + \text{Loss}(f_i)\right) \quad (6.11)$$

and its value can be calculated by integrating this uniform distribution up to the current value of $\text{Loss}(CA_j, f_i)$ (see Eq. (6.5)). To calculate the probability in Eq. (6.10), the facts that performance of the system under consideration is just dependent on the faults and the corrective operations, and the losses are defined by the fault and operations and their benefits (see Eqs. (6.6) and (6.7)), have been used.

6.5 An illustrative example

In this section, the application of the proposed decision support system is demonstrated on the CSTH benchmark. For this study, two different faults are introduced to the benchmark. Fault f_1 is a drift in temperature sensor and fault f_2 is an actuator fault in hot water valve (see Fig. B.1). The fault f_1 affects the temperature inside the tank and fault f_2 affects the temperature and the level of the liquid inside the stirred tank. As a result, the quality of the final product is affected due to the non-optimal condition for the reaction.

In the off-line training step, the data with respect to the fault-free and faulty operations of the process are collected and the corresponding statistical models are derived. The results are shown in Table 6.1. The results are further utilized in the on-line step to calculate the probability of each individual fault and to propose the suitable maintenance operation.

From the process knowledge a list of maintenance operations is defined for each fault. The losses due to the faults and maintenance actions related to each fault is shown in Table 6.2. Moreover, the fixed costs of each maintenance operation and the parameters α_{ij}, which are basically the benefit of performing CA_j in case of fault i, are represented in Table 6.3. It is assumed that the value of the production till to the next planned maintenance of the process is 10000 units.

Due to a drift fault in the temperature sensor, the controller changes the actuator signal in steam valve to compensate the false reading of the sensor. Therefore, one possible solution to the sensor fault in this case could be adjusting the steam valve's actuated signal to compensate this false reading of the sensor. Another solution could be replacing the sensor which is subjected to higher cost but it provides the possibility to completely remedy the effects of fault on the product quality. Alternatively, one can wait till next planned shutdown of the process to replace the sensor.

In the case of the fault in hot water valve, the flow of hot water into the stirred tank will reduce. As a result, the level of water and its temperature will decrease. By increasing the cold water flow and

Table 6.2: Maintenance operations with respect to each fault and losses due to each fault

Fault	Losses	Maintenance operation
f_1	3000	Adjust the steam valve
		Replace the sensor
		Do nothing
f_2	8000	Adjust the cold water and steam valves
		Reduce the production
		Replace the valve
		Do nothing

Table 6.3: Maintenance operations, their costs and benefits

Sig.	Maintenance action	Fix costs	α_{ij}	
			f_1	f_2
CA_1	Adjust the steam valve	100	0.75	0
CA_2	Replace the sensor	200	1	0
CA_3	Adjust the cold water and steam valves	200	0	0.75
CA_4	Reduce the production	300	0	0.8
CA_5	Replace the valve	1000	0	1
CA_6	Do nothing	0	0	0

increasing the steam flow, the optimal condition for the reaction will be assured. But it causes higher energy consumption. Another solution besides doing nothing or replacing the valve, might be reducing the production rate by decreasing the cold water flow which results in less turnover.

To simulate this concept on the CSTH benchmark, a scenario is considered where the process is working in normal operating condition at the beginning and then the faults are appearing in the system. The probability of the three different episodes, namely process operation is normal or $p(\mathbf{x} \in N)$, is subjected to fault f_1 or $p(\mathbf{x} \in f_1)$ and is subjected to fault f_2 or $p(\mathbf{x} \in f_2)$ for this scenario are shown in Fig. 6.2. Using these probabilities, the *a priori* knowledge is obtained in the training phase in Table 6.1 and information about maintenance operations in Tables 6.2 and 6.3, the suitable operation for each sample of measurement is calculated using Eq. (6.9) and plotted in Fig. 6.3. The plot in Fig. 6.3 shows that, given the probability of each event in the plant, which maintenance operation is the most suitable one in terms of lowest costs and highest impacts on the quality of the production.

To represent the results of decision support system in more details, they are shown again in Figs. 6.4 and 6.5 for selected intervals. It can be seen in the first episode, where the probability of normal operation

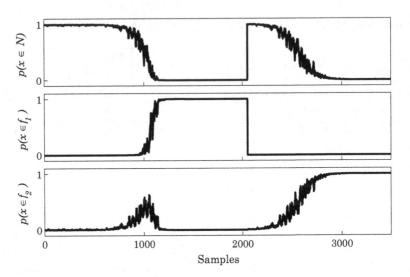

Figure 6.2: Probability of the faults

is high, the best maintenance action is doing nothing. After that, when the probability of fault f_2 is increasing, the decision system

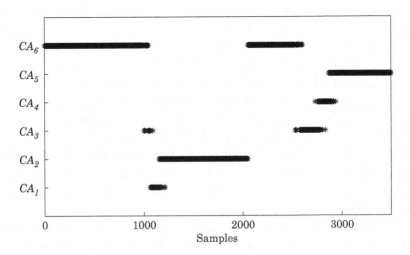

Figure 6.3: Results of corrective operation generation

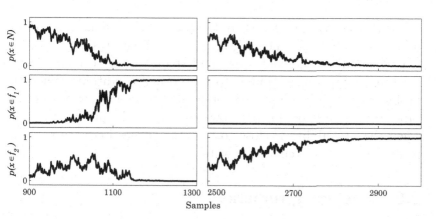

Figure 6.4: Probability of the faults (zoomed)

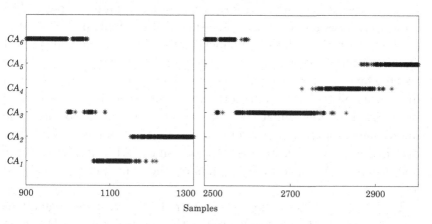

Figure 6.5: Results of corrective action generation (zoomed)

suggests to perform maintenance operation CA_3 which is adjusting the cold and steam valves, since it can improve the product quality with the lowest costs. Furthermore, when the probability of fault f_1 is increasing, the best operation is CA_1 and when the probability is high enough and adjusting the valve is inducing high risk, stopping the process and repairing the sensor is going to be the best operation.

For the second episode, where the process is running from normal operation to fault f_2, the suggestion of the decision support system is first to adjust the cold water and steam valves to maintain the level and temperature similar to the normal condition. When the probability of f_2 increases it is preferable to reduce the production rate, to have high quality but low quantity of the product. Finally, when the probability is close to one, the suggestion is to stop the plant and repair or replace the valve.

6.6 Concluding remarks

In this section, a novel decision support system is proposed which provides the best maintenance operation once a fault happens. The final decision is based on the assessment of the fault impact on the product quality as well as its financial consequences. The fault impact analysis is obtained by monitoring the subspace of the process variable space which has correlation to the product quality and is described in a probability form. Furthermore, this probability is combined by a risk function which characterizes the risk of performing the maintenance operation assuming a certain fault is happening. The best maintenance operation is selected using MAP criterion as those operations which have high impacts on the plant performance and low costs.

The ability of this method to consider the expert knowledge about certain events in the process as well as the uncertainties in FD system is of great importance from the practical point of view. Moreover, the performance of the presented decision support system can be improved in practice by iterative learning of the performance behavior after performing a certain corrective operations. That means the main parameters in the decision system e.g. fault models ($p(\mathbf{x}|f)$) and benefit of performing a corrective action (α_{ij}), can be easily relearned, which is a superiority compared to the existing methods.

7 Application and benchmark study

The purpose of this chapter is to demonstrate the applications of the proposed techniques in this thesis on the industrial benchmark plants and discuss their performance and effectiveness. Although a simulation example is given at the end of each chapter to show the ability of the methods, it is interesting to see the application of the proposed methods in real industrial processes. To this aim, the algorithms developed in this thesis have been applied to a laboratory setup of continuous stirred tank heater (CSTH) and the dryer section in a paper and board machine. Before starting the discussion about the application of the methods on the processes, a brief description about each process is given.

The main idea of the proposed methods is to design the monitoring and diagnosis scheme using the process historical data for nonlinear multimode processes. Therefore, only the sufficient physical descriptions of the process are given to show the overall dynamics and nonlinear behavior of the plant. However, the detailed model of the processes can be found in given references.

7.1 Laboratory CSTH setup

CSTH plants are widely used in chemical industry to ensure optimal conditions for chemical reactions. Inside the CSTH a certain temperature and level of reactants are being held, that define the operating point under which an optimal reaction is possible. The laboratory sized CSTH considered here is a RT 682 CSTH demonstrator plant

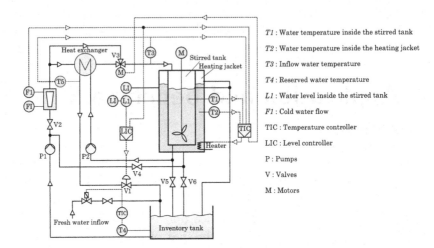

Figure 7.1: Piping and instrumentation diagram of CSTH plant

manufactured by G.U.N.T. Geraetebau GmbH Hamburg[1], available at
the institute of Automatic Control and Complex Systems, University
of Duisburg-Essen. It uses water as reactants and is structurally
depicted in Fig. 7.1. The plant's main component is a tank in which a
certain amount of the preheated reactant is mixed, heated further and
held at a certain temperature. The plant has a steady through flow
of reactants which is controlled via a hand valve on the inflow side.
An actuated valve controls the outflow and can thus be used as an actuator for level control. The temperature inside the tank is raised via
a heater in the surrounding water filled heating jacket, whose power
can be controlled. By heating the jacket, a steady heat flow through
the tank's walls and into the reactant will increase the temperature
inside the stirred tank. The inflowing reactant is pre-heated from the
outflowing product via a heat exchanger, so that it is only $5 - 10\,°C$
colder than the desired tank temperature.

---

[1]see www.gunt.de for details.

The dynamics of the plant are described by mass balance and energy balance equations:

$$h_{tank}(t_1) = \int_{t_0}^{t_1} \frac{1}{A \cdot \rho}(\dot{m}_{in}(t) - \dot{m}_{out}(t))dt + h_{tank}(t_0), \qquad (7.1)$$

and

$$T_{tank}(t_1) = \int_{t_0}^{t_1} \frac{1}{c_p \cdot m(t)}(Q_{in}(t) - Q_{out}(t))dt + T_{tank}(t_0), \qquad (7.2)$$

where h_{tank} is the level of water inside the tank [m], $\dot{m}_{in}, \dot{m}_{out}$ are in- and outflowing water mass flow rates [kg/s] , T_{tank} is the temperature of water in the tank [K], Q_{in}, Q_{out} are the rates of heat flow in and out of the tank [W] and A, ρ, c_p, m are the cross section of the cylindrical tank [m^2], density of water [kg/m^3], specific heat capacity of water [J/(kg.K)] and the mass of water in the tank [kg], respectively.

The level and temperature are directly measurable via sensors, as well as the mass flow of water into the tank $\dot{m}_{in}$. The mass flow out of the tank $\dot{m}_{out}$ is directly manipulated via a pneumatic valve. The mass of water in the tank m, that defines the relation between the enthalpy and the temperature of the water in Eq. (7.2), is supposed to be constant in an operating point, so that its value can be put in front of the integral. The heat flow into the tank Q_{in} is a nonlinear function of P_{heater} which is the controllable power of the heater [W], h_{hj} the water level in the heating jacket [m] and the level of the water in the tank. Furthermore, the preheated inflowing reactant and the outflowing product provide an additional heat flow in and out of the tank, Q_{infl} and Q_{outfl} respectively, which are functions of the mass flows and the temperature of those masses. These are considered to be constant in an operating point as well as the heat flows. To describe heat losses due to non-optimal insulation, the heat flow out of the system, Q_{out} was introduced in which all non-optimal effects

Table 7.1: Values defining plant operating points

values	$\mathcal{M}_1$	$\mathcal{M}_2$	$\mathcal{M}_3$
level L1 [cm]	10	12	18
level heating jacket [cm]	20	20	20
temperature T1 [°C]	50	45	50
through flow F1 [L/h]	105	105	105

can be lumped. Taking all this into consideration, Eqs. (7.1) and (7.2) become

$$h_{tank}(t_1) = \frac{1}{A \cdot \rho} \int_{t_0}^{t_1} (\dot{m}_{in}(t) - \dot{m}_{out}(t))dt + h_{tank}(t_0)$$

$$T_{tank}(t_1) = \frac{1}{c_p \cdot m} \int_{t_0}^{t_1} (Q_{in}(P_{heater}(t), h_{hj}, h_{tank}(t))$$

$$- Q_{out}(t) + Q_{infl}(t) - Q_{outfl}(t))dt + T_{tank}(t_0) \qquad (7.3)$$

From this, the overall dynamic behavior of the two subsystems for level and temperature can be seen as well as the nonlinearities and operating point dependencies of the system model.

7.2 Multimode fault diagnosis in CSTH testbed

The application of multimode FD technique for static systems, proposed in Chapter 3, is studied in this section. To carry out this study, three different operating points, as shown in Table 7.1, are chosen for the CSTH testbed. For each operating point 500 samples were used to train the model using Algorithm 2. The level of the reactant inside the stirred tank, $L1$, is considered as quality variable, Y and the remaining variables are considered as input variables $\mathbf{X}$.

For validation purpose, faulty operation is induced by scaling the signal given to level controlling valve $V1$ behind the controller output

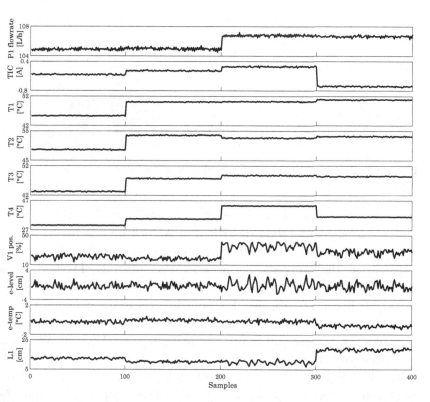

Figure 7.2: Process variables for CSTH plant

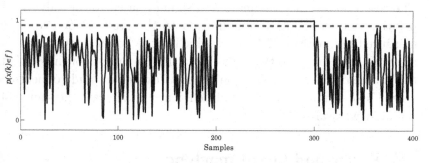

Figure 7.3: Result of fault detection using proposed static approach

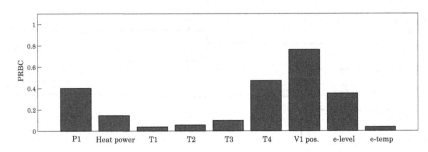

Figure 7.4: Probabilistic reconstruction based contribution plot

In on-line monitoring step, the process variables $\mathbf{X}$ are measured and the proposed method described in Algorithm 2 is implemented for monitoring. The fault detection index $p(\mathbf{x}(k) \in f)$ is depicted in Fig. 7.3 over the faulty and normal operation intervals. The horizontal dashed line represents the threshold with 97% confidence level for $\alpha = 0.03$. It can be seen that the fault is successfully detected within the confidence level. Figure 7.2 shows the run of all measurable plant variables for normal and faulty operation in $\mathcal{M}_1$, $\mathcal{M}_2$ and $\mathcal{M}_3$ each consisting of 100 samples.

To identify the possible source of malfunction in the system, the probabilistic contribution analysis method, developed in Chapter 5 is applied after detection of the fault. The result is shown in Fig. 7.4. The result shows that the measured signal related to the valve $V1$'s position has the highest contribution to the fault. Moreover, the measured temperature $T4$ signal which is the liquid temperature in reservoir contributes to the fault. From that, it can be concluded that the source of the fault might be the heat exchanger and connecting valve $V1$ or pump, which is confirmed, since in this case the source of the fault was valve $V1$.

7.3 Paper and board machine

Papers are nowadays used in a variety of applications, ranging from simple printing to some advanced building materials. Depending on their application they are produced in different grades and specifications. Basic properties which are particularly important and differ from one grade to another, are as follows [49]:

- **Basis weight (grammage)** $[g/m^2]$: total mass of $1m^2$ paper considering all of its components, e.g. fiber, filler, water, etc.

- **Bulk** $[kg/m^3]$: which is basically the density of the paper. It strongly affects the mechanical properties of the paper such as bending, stiffness, etc.

- **Formation** $[g/m^2]$: represents the basis weight variation in the paper and distribution of the fibers in the paper.

- **Brightness** [%]: represents the amount of light absorption in the paper and is an important characteristic of paper for printing applications.

In Germany, the total production of paper and paperboard according to the statistics in 2006 was 33.320 million tons, with a turnover of more than 35 billion euros and approximately, 1700 different enterprises are active in this field [35].

The task of a paper machine is to make paper from pulps and other raw materials. A typical paper machine mainly consists of three sections [49]:

- **Wire section:** in wire section, fiber web is formed from the pulp suspension entered from headbox. The main objective of wire section is to distribute the suspension into the web of the width of whole machine and deliver desired consistency, thickness and speed to the wire pulp. The pulp suspension entering the wire section contains only about 1% of fiber. As the web travels through the wire section, the liquid content is drained by gravitational forces or suction. The dry content of web at the end of wire section is about 20%.

- **Press section:** the purpose of press section is to reduce the water content in the fiber web as much as possible by compression of the

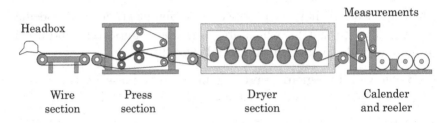

Figure 7.5: Diagram of a typical board machine

web between rotating steel rolls. It also increases the mechanical strength of the web and prevents web breaks in production line. In press section the dry content of web will be increased form 20% to about 50%.

- **Dryer section:** in dryer section the solid dry content of web is increased up to 90% by evaporation. The heat is transfered from steam heated large hollow metal cylinder to the paper web and evaporates the water. To improve some physical properties of the paper e.g. printability, usually additional coating materials are sprayed on the paper in dryer section. Finally after dryer section, the paper is wound up on a roll and removed from the machine.

The diagram of a typical paper machine is depicted in Fig. 7.5. The dryer section in a paper machine is the most important part in term of energy consumption. Around 2/3 of energy consumed in a paper machine is used in dryer section to reduce the moisture content in the paper web [99]. Moreover, the major physical properties of final paper product such as elasticity, twist and stiffness are affected by the moisture content as well as performance of the dryer section's operation. These make the moisture one of the most important quality variables in the paper making industry from operational and economical points of view. During the production, the moisture content of the paper web is measured and controlled to maintain a uniform moisture distribution. As a result, an effective monitoring

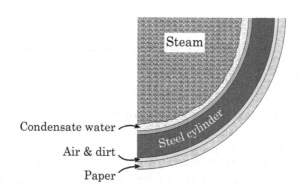

Figure 7.6: Steam-filled cylinder in dryer section

and diagnosis tool for dryer section can strongly improve the process performance and reduce the production costs.

7.3.1 Dryer section

Before entering the dryer section, about 50% of the paper web consists of water and the major task of dryer section is to reduce water content to 5-10%. The most common way to remove the moisture from paper in dryer section is to transfer the heat from steam-filled cylinder to the paper surface. Inside the cylinder, the thermal energy in saturated steam is transferred to the steel shell and used to evaporate the water content in the sheet. After loosing its thermal energy, the steam condensates in the cylinder. The condensate water inside the cylinder will reduce the heat transfer rate. Moreover, the mechanical energy required to rotate the cylinder will increase due to the condensate water. Therefore, it is fed out through a siphon. The cross section of a typical cylinder in dryer section is shown in Fig. 7.6.

The cylinders in dryer section are arranged in different groups. The steam pressure in each group is controlled such that the desired pressure profile is obtained in entire dryer section. Furthermore, the moisture content in the paper is controlled by changing the steam pressure. The moisture is measured using a scanners in the measurement unit before calender section (see Fig. 7.5). A typical moisture

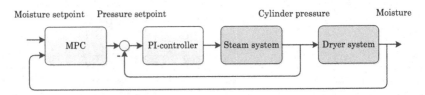

Figure 7.7: Diagram of moisture control loop

control loop is shown in Fig. 7.7 [99]. The moisture control loop usually has a cascade structure, where the inner control loop controls the group's steam pressure via a proportional-integral (PI) controller, and the outer loop is a model predictive controller (MPC) responsible for moisture control by providing a set-point to pressure control loop

7.3.2 Steam-filled cylinder model

To describe the physical behavior of the dryer section and study its dynamics and nonlinear properties, in this section the model of a steam heated cylinder is derived based on the mass and energy balances. For comprehensive modeling of the dryer section in a paper machine, the readers are referred to [7, 69, 99] and the references therein.

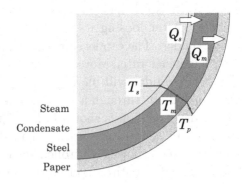

Figure 7.8: Thermal energy transfer from steam to the paper web [99]

The flow of thermal energy from steam to the paper web is shown in Fig. 7.8. The temperature profile is also depicted therein. Inside the cylinder, there is a constant steam temperature and the temperature profile gradually decreases with different gradient due to different heat transfer coefficients of the materials.

Considering the mass balance equations for the steam and water inside the cylinder, it turns out that

$$\frac{\mathrm{d}(\rho_s V_s)}{\mathrm{d}t} = \dot{m}_s - \dot{m}_c,$$
$$\frac{\mathrm{d}(\rho_w V_w)}{\mathrm{d}t} = \dot{m}_c - \dot{m}_w, \tag{7.4}$$

where ρ_s and ρ_w are densities of the steam and water [kg/m^3], V_s and V_w volume of steam and water inside the cylinder [m^3] , $\dot{m}_s, \dot{m}_c$ and $\dot{m}_w$ are mass flow rates of steam into the cylinder, condensation rate and mass flow rate of water out of cylinder [kg/s], respectively. Using the notations h_s and h_w for steam and water enthalpies [J/kg], u_s and u_w as specific internal energies of steam and water [J/kg] and C_{pm} for specific heat capacity of the cylinder [J/(kg.K)], respectively, the energy balance equations can be written as

$$\frac{\mathrm{d}(\rho_s V_s u_s)}{\mathrm{d}t} = \dot{m}_s h_s - \dot{m}_c h_s,$$
$$\frac{\mathrm{d}(\rho_w V_w u_w)}{\mathrm{d}t} = \dot{m}_c h_s - \dot{m}_w h_w - Q_s,$$
$$\frac{\mathrm{d}(m C_{pm} T_m)}{\mathrm{d}t} = Q_s - Q_m, \tag{7.5}$$

where m is the mass of the cylinder [kg] and Q_s and Q_m are thermal power [W] transferred from water to the metal shell and from metal shell to paper web, respectively, which can be described as follows

$$Q_s = \kappa_s a_c (T_s - T_m),$$
$$Q_m = \kappa_m a_c (T_m - T_p), \tag{7.6}$$

where κ_s and κ_m are heat transfer coefficients from steam to the metal shell and from metal shell to the paper web [W/(m^2.K)] and a_c is the cylinder area [m^2].

Considering the dynamics of paper drying process, it has been shown in [99] that the mass and energy balances in paper web can be described as

$$\frac{\mathrm{d}(uga_{xy})}{\mathrm{dt}} = d_y v_x g u_{in} - a_{xy} \dot{m}_{evap} - d_y v_x g u,$$

$$\frac{\mathrm{d}(g(u+1)a_{xy}C_{pp}T_p)}{\mathrm{dt}} = Q_m + d_y v_x g(1 + u_{in})C_{pp}T_{p,in}$$
$$- a_{xy}\dot{m}_{evap}\Delta H - d_y v_x g(1+u)C_{pp}T_p \quad (7.7)$$

where g is the dry basis weight [kg/m^2], u is the moisture ratio [kg moisture/kg fiber], a_{xy} is the area of the cylinder which is in contact with paper web [m^2], C_{pp} is the specific heat capacity of the paper web [J/(kg.K)], d_y is the paper width [m], v_x is the speed of the paper web [m/s], $\dot{m}_{evap}$ is the evaporation rate [kg/(m^{2}s)] and ΔH is the amount of energy required to evaporate the water from the paper surface [J/kg].

The first principle models of the cylinder shown in Eqs. (7.4) to (7.6) and the moisture evaporation in Eq. (7.7) are dynamic nonlinear models. Although, assuming some conditions ([99]) the model can be described by a set of linear differential equations, the linearized model is valid around a specific operating point and the model parameters may change from one to another operating point.

The dryer section in a paper machine, consists of several different components, for instance pumps, valves and hydraulic devices. The components are subjected to different kinds of faults and malfunctions. These faults strongly affect the product quality as well as availability of the process. Web breaks due to disengagement of cylinders are inevitable in the dryer section and cause unplanned shutdown of process. Leakage in the pipes is the main source of deviations in the moisture and leads to waste of energy in dryer section. Valve stiction, oil leakage, pump and drive overheating are other types of common malfunctions in dryer section [57]. Numerous methods for detection and analysis of the fault in paper machine have been developed and published. Most of them are designed to detect a fault in a certain component in the machine. More details about the approaches can be found in [3, 8, 18, 19, 33, 53, 54, 71, 100, 104].

In the remaining part of this chapter, the methods developed in preceding chapters are demonstrated using the process measurements obtained from the described board machine.

7.4 Application of static FD method on dryer section

The main aim of this section is to demonstrate the application of the proposed method in Chapter 3 on the dryer section of a board machine to detect those faults which are affecting the product quality. In this demonstrative example, the moisture measurement is considered as the product quality measurement, since it is mainly characterized by the performance of the dryer section. Although different faults exist in dryer section, not all of them affect the product quality. For instance, the scatter plot of moisture versus the main pressure is given in Fig. 7.9. The data points denoted by N_1 and N_2 represent the normal operating data corresponding to two different modes. The area denoted by f_1 and f_2 represent two different fault scenarios. It can be seen that, in the case of f_1 the moisture is not affected and remains inside an acceptable range. However, the fault f_2 has a strong impact on the moisture and consequently leads to production loss. In such cases, early detection of any deviation in the product quality is of great importance and can remedy the financial consequences of failure. For such processes which are working on different operating regimes, the fault detection algorithm should be able to distinguish between normal operating modes and faulty modes and at the same time be capable to assess their impacts on the quality variables.

For this case study, the processes variables corresponding to the dryer section are considered. To cut down the complexity of the problem, using the variable selection methods, e.g. cross validation test [121] or structural modeling approaches [4], the dimension of the data is reduced to the process variables which are affecting the

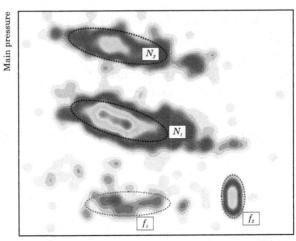

Figure 7.9: Scatter plot of moisture vs. pressure

moisture. As the result, 10 process variables are selected to be used to carry out this study. The chosen variables are steam pressure measurements in different cylinders, production rate, basis weight of the paper web and the machine speed.

The multimode FD technique described in Algorithm 2 is further employed to design the fault detection scheme. For training purpose, the data with respect to three different paper grades are collected These paper grades are characterized by different basis weight, thickness and moisture specification. The data are pre-processed, i.e. the transient behavior presented in the training data is excluded, since the method is best suited for stationary processes. Furthermore, the EM algorithm presented in Eqs. (3.22) to (3.29) is used to obtain the parameters of the mixture model, namely $\mathbf{M}_i, \boldsymbol{\mu}_{x,i}, \boldsymbol{\mu}_{y,i}, \boldsymbol{\Sigma}_{xx,i}, \boldsymbol{\Sigma}_{yy,i}$ for $i = 1, 2, 3$. These parameters are used to design the monitoring scheme in forthcoming steps.

To simulate the on-line monitoring step, again data corresponding to those three different paper grades are collected from another period of time. In addition, a faulty episode which caused unplanned shutdown of the production line is considered. The measurements used for monitoring step are plotted in Fig. 7.10. At the beginning, the process is running in its normal operating modes. In sample 2635 a fault happens which is followed by a plant shutdown at sample 2741. As explained in Algorithm 2, in on-line step when a new sample of measurement $\mathbf{x}(k)$ is available the probabilities that the sample is belonging to the modes, i.e. $p(\mathbf{x}(k) \in \mathcal{M}_i)$ are calculated for $i = 1, 2, 3$. These probabilities are calculated using Bayesian inference strategy presented in Eq. (3.32) and plotted in Fig. 7.11. Furthermore, the probabilities that the sample is faulty, assuming that it is generated in mode $\mathcal{M}_i$, i.e. $p(\mathbf{x}(k) \in f | \mathbf{x}(k) \in \mathcal{M}_i)$ are obtained by Eq. (3.33) and plotted in Fig. 7.12 for different modes or paper grades.

Finally, the global fault detection index in Eq. (3.31), which demonstrates the probability that a quality related fault happens in the system is calculated by combining the two aforementioned probabilities in Figs. 7.11 and 7.12 and shown in Fig. 7.13. The red dashed line represents the 95% confidence level. As it is shown, the fault is successfully detected at sample 2636 with one sample detection delay. The false alarm rate and missed detection rate are 0.61% and 1.87% respectively in this case. In order to provide a comparison between the proposed methods for multimode processes and the standard approaches, the simulation is carried out using the modified PLS approach [128], described in Section 3.2. Using the same training data and considering the identical episodes for on-line simulation, the logarithm of calculated $T_{\hat{x}}^2$ index, which is used to detect quality related faults, is shown in Fig. 7.14. The false alarm and missed detection rates are 3.26% and 93.46% respectively. As it can be seen from the results, using the multimode approach, the performance of the fault detection scheme has been considerably improved. Especially considering the missed detection rate, it can be concluded that the fault is not detected using the standard multivariate technique.

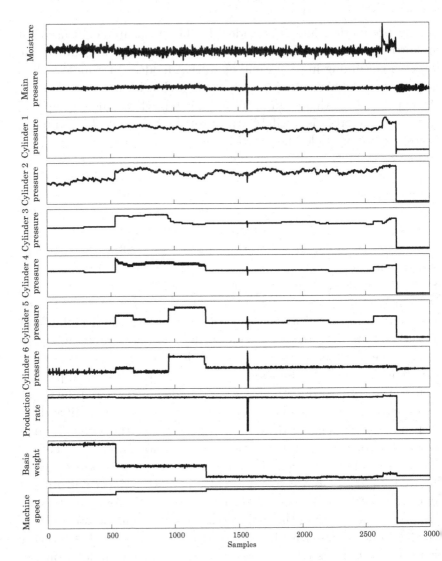

Figure 7.10: On-line measurements for the case study

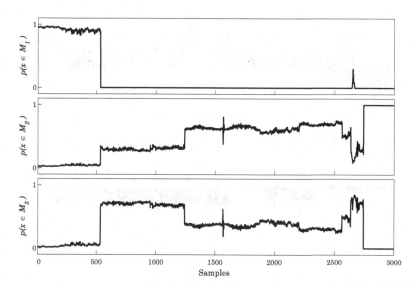

Figure 7.11: Probability that the samples are generated in mode $\mathcal{M}_k$ for $k = 1, 2, 3$

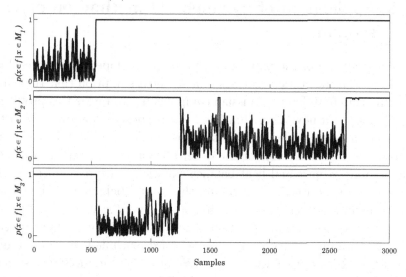

Figure 7.12: Probability that the samples are faulty given that they are generated in mode $\mathcal{M}_k$ for $k = 1, 2, 3$

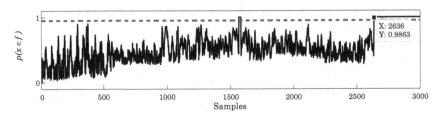

Figure 7.13: Probability that the samples are faulty

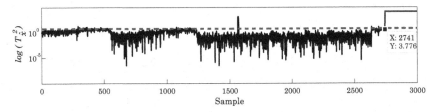

Figure 7.14: Results of FD using MPLS method

7.5 Application of dynamic FD method on dryer section

The application of the multimode approach developed in Chapter 3 is restricted to the static processes, due to the assumption that the data is a mixture of different Gaussian components. To tackle this problem, in Chapter 4 a new algorithm has been proposed to address the FD problem in multimode dynamic systems. In Section 7.3.2, it has been shown that the behavior of the dryer section can be represented by the nonlinear mass and heat balance equations with dynamics shown in Eqs. (7.5) and (7.7). Therefore, the overall behavior of the dryer section can be described as a dynamic nonlinear system.

In this section, the application of proposed multimode fault detection method on the dryer section of a board machine is demonstrated considering its dynamics. Same as Section 7.4 the moisture is considered as the product quality variable. The same set of measurements is used as the process variables which are describing the moisture and its variations.

For this case study, a scenario is considered where the process is operating in two different modes and producing two different paper grades. Using the available historical data, the off-line training is carried out according to the steps shown in Eqs. (4.16) to (4.21) to design the observers in Eq. (4.22).

For on-line monitoring purpose, data corresponding to those two paper grades are collected, followed by a fault and an unplanned shutdown. The plot of data is shown in Fig. 7.15 and the effect of fault on the quality variable, namely moisture, can be seen therein. In this simulation step, when a new sample of measurement $\mathbf{d}(k)$ is available, the probabilities that the sample is generated in different modes, $p(\mathbf{d}(k) \in \mathcal{M}_i)$ are calculated using Eq. (4.24) and shown in Fig. 7.16. Using the identified observers in off-line training step and the observed signal $\mathbf{d}(k)$, the test signals $J_i(k)$ in Eq. (4.25) are built and based on it the probabilities $p(\mathbf{d}(k) \in f | \mathbf{d}(k) \in \mathcal{M}_i)$ are calculated by Eq. (4.28) for these two different modes. These probabilities are shown in Fig. 7.17.

Finally, by combining the probabilities that the sample is faulty under the assumption that it belongs to mode $\mathcal{M}_i$ and the hypothesis that it is generated in this mode, the global fault detection index is calculated by using Eq. (4.27). The result shown in Fig. 7.18, represents the successful detection of the fault.

To compare the performance of this algorithm with the static multimode fault detection algorithm developed in Chapter 3, the result obtained by static method for the same scenario is shown in Fig. 7.19. The superior performance of the dynamic approach, especially in the instants where the operating point is changing, can be seen by comparing Figs. 7.18 and 7.19. The false alarms are reduced by incorporating the dynamic model in mixture modeling step. Although the computation complexity of the dynamic FD method is higher than the static method, its performance and accuracy make it extremely suitable for monitoring of dynamic systems.

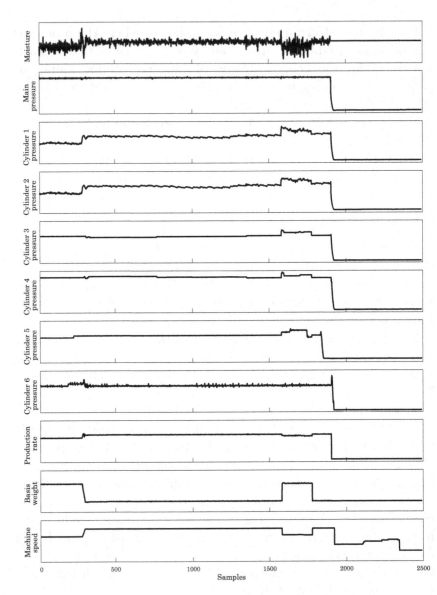

Figure 7.15: Measurements for the dynamic case study

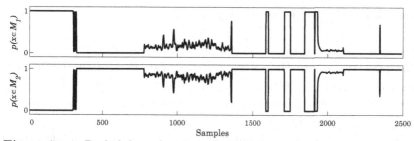

Figure 7.16: Probability of each modes for dynamic case study

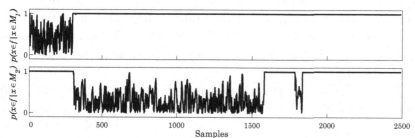

Figure 7.17: Probability of the fault given the modes for the dynamic case study

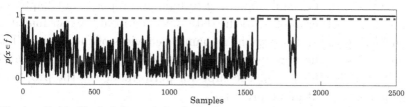

Figure 7.18: Probability of the fault for the dynamic case study

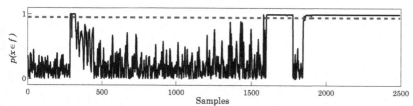

Figure 7.19: Results of the static FD technique using the measurements of the dynamic case study

7.6 Concluding remarks

The main objective of this chapter is to demonstrate the performance and effectiveness of the developed methods in this thesis through their application on industrial plants. For this purpose, a laboratory scale CSTH plant and a large scale industrial plant, namely a paper machine, are considered.

In the first part of this chapter, the CSTH testbed is described and its nonlinear behavior and set-point dependencies are presented. Furthermore, the multimode FD technique developed in Chapter 3 is used to detect a fault in a valve. The probabilistic method proposed in Chapter 5 is further used to identify the possible source of the fault in the plant.

The second part of this chapter is dedicated to application of the multimode fault detection techniques on dryer section of a paper machine. The dryer section is a realistic example of a large-scale process, which consists of several hundreds measurements and control loops. To this end, firstly a brief description of a typical paper machine is given. The multimode FD techniques proposed in Chapters 3 and 4, are further applied to the data obtained from a paper machine considering the stationary and dynamic behaviors of the system. The efficiencies of the new methods are also compared with the standard methods which show the superior performance of the proposed methods for multimode processes.

8 Summary

This thesis is aimed to contribute to the development of new data-driven techniques for fault detection, diagnosis and treatment in complex industrial processes. In Chapter 1, the basic idea of FDD systems is explained, their historical evolution and major developments in recent years are given. In this chapter it has been identified that the complexity of modern industrial processes puts restrictions on the application of the model-based FDD methods on large-scale processes. In addition to that, an alternative approach which makes full use of process historical data to design the FDD system is introduced. Moreover the drawbacks of the classical data-driven methods for designing the FD systems which limit their applications to LTI systems have been discussed. The main objective of this thesis is to address the above mentioned obstacles and propose alternative approaches to deal with these circumstances. The main results and conclusion of this thesis is summarized in this chapter.

To begin with, an overview of different model-based FDD techniques is given in Chapter 2. Observer-based residual generation, parity space-based residual generation and their interconnections and relations are elaborated therein. These techniques are based on deriving the mathematical model of the process under consideration through rigorous approaches, e.g. via first principles. An alternative approach, i.e. data-driven technique, which uses the process historical data to abstract the diagnosis information is described in the remaining part of this chapter. The most popular data-driven techniques, such as multivariate statistical process monitoring techniques and subspace-based design of FDD system are elaborated therein.

The available data-driven techniques are based on the assumption that the process under consideration is a linear system or linearized around a single operating point. However, in reality it might not be valid, since the processes are nonlinear and working on different operating points due to different product specifications and restrictions. The fault detection system needs to be robust against such changes in the process. These issues are addressed in Chapters 3 and 4 and new techniques are proposed for design of FD systems in nonlinear multimode static and dynamic processes. These techniques are based on the efficient identification of the mixture model from the process data using EM algorithm in the off-line modeling step and using a Bayesian inference technique for fault detection in on-line monitoring step.

After successful detection of faults and malfunctions in the system, the next steps are to identify the possible sources of the fault and perform the suitable corrective maintenance operation. To this aim, in Chapter 5, a new method is proposed which addresses the problem of fault isolation in multimode processes. This method combines the information about the impacts of the fault on the process variables and the current operating mode of the process to rank the process variables with respect to the fault impacts. The results of this method can be used as a guideline for process engineer to recognize the root-causes of the faults in the system. Furthermore, in Chapter 6 a novel decision support system is presented which integrates the information about possible faults in the system and their economic assessment to provide a list of proper corrective maintenance operations. A Bayesian technique is used to obtain this information in terms of probability of the different faults in the system. This probability is further combined with a cost-benefit analysis to find the optimal maintenance operation

Chapter 7 is devoted to study the performance of the proposed data-driven techniques on industrial applications. A laboratory scale CSTH plant and the dryer section of a board machine are considered to carry out this study. The proposed methods are applied on the above mentioned examples and different aspects of their application are described in this chapter. The results of different techniques are compared and their performance and effectiveness are discussed.

Although the results of this thesis provide solutions to some obstacles, there are still some issues that need to be addressed. In this thesis, the nonlinear systems are assumed as PWA systems. However, many industrial applications follow certain class of nonlinear behaviors e.g. Hammerstein [116] or Wiener [118] systems. Extension of the proposed methods to these classes of nonlinear systems is an issue that requires more attention. Reducing the numerical complexity of the methods using other efficient mixture modeling tools is another interesting point that can be addressed in the future works.

Appendices

A Expectation-maximization algorithm

Expectation-maximization (EM) is an iterative algorithm used for maximum likelihood estimation (MLE) in the situations where the available data are incomplete. EM algorithm has been proposed in [27] for the first time, and its properties and applications have been presented. Each iteration of EM algorithm consists of two steps; expectation or E-step and maximization or M-step, where the conditional expectation of the log-likelihood of data is calculated in E-step and parameter estimation update is carried out in M-step. This section provides a brief overview of EM algorithm, its derivation and extensions.

A.1 Derivation of EM algorithm

The basic idea of EM algorithm is to solve a complete data MLE problem, given incomplete data set. Therefore, it is necessary to begin with a brief discussion of MLE.

A.1.1 Maximum likelihood estimation

MLE method tries to estimate the unknown parameters Θ such that the observed measurements $\mathcal{D} = \{d_1, d_2, \cdots, d_N\}$ become as likely as possible. In other words, the unknown parameters are estimated as

$$\hat{\Theta}_{MLE} = \arg\max_{\Theta}\{p(\mathcal{D}; \Theta)\}. \tag{A.1}$$

Since the logarithm is a strictly increasing function, it is often convenient to represent the MLE problem in the following form

$$\hat{\Theta}_{MLE} = \arg\max_{\Theta}\{\log p(\mathcal{D}; \Theta)\}. \tag{A.2}$$

The log-likelihood function for Θ formed from the observed data $\mathcal{D}$ is defined by

$$L(\mathcal{D}; \Theta) = \log p(\mathcal{D}; \Theta) \tag{A.3}$$

The estimate $\hat{\Theta}$ of Θ is obtained by the solution of MLE as

$$\frac{\partial L(\mathcal{D}; \Theta)}{\partial \Theta} = 0. \tag{A.4}$$

The solution in Eq. (A.2) can be obtained analytically or using standard methods such as Newton-type methods.

A.1.2 Expectation-maximization

As mentioned earlier, the EM algorithm is used to solve MLE in the case that the given observations are incomplete. The basic idea in EM is to consider the joint log-likelihood function of complete data set $\mathcal{C} = \{\mathcal{D}, \mathcal{Z}\}$, where $\mathcal{D}$ denotes the given incomplete observations and $\mathcal{Z}$ represents the hidden or latent variables,

$$L(\mathcal{Z}, \mathcal{D}; \Theta) = \log p(\mathcal{Z}, \mathcal{D}; \Theta) \tag{A.5}$$

Following the definition of conditional probability

$$p(\mathcal{Z}|\mathcal{D}; \Theta) = \frac{p(\mathcal{Z}, \mathcal{D}; \Theta)}{p(\mathcal{D}; \Theta)}, \tag{A.6}$$

it turns out that

$$\log p(\mathcal{D}; \Theta) = \log p(\mathcal{Z}, \mathcal{D}; \Theta) - \log p(\mathcal{Z}|\mathcal{D}; \Theta). \tag{A.7}$$

Multiplying and summing both sides of Eq. (A.7) with respect to $p(\mathcal{Z}|\mathcal{D}; \Theta_k)$, where Θ_k is the estimated value of Θ in k^{th} iteration, leads to

$$\log p(\mathcal{D}; \Theta) = \sum_{\mathcal{Z}} \log p(\mathcal{Z}, \mathcal{D}; \Theta) p(\mathcal{Z}|\mathcal{D}; \Theta_k)$$
$$- \sum_{\mathcal{Z}} \log p(\mathcal{Z}|\mathcal{D}; \Theta) p(\mathcal{Z}|\mathcal{D}; \Theta_k)$$
$$= E\left\{\log p(\mathcal{Z}, \mathcal{D}; \Theta)|\mathcal{D}; \Theta_k\right\}$$
$$- E\left\{\log p(\mathcal{Z}|\mathcal{D}; \Theta)|\mathcal{D}; \Theta_k\right\}$$
$$= \mathcal{Q}(\Theta; \Theta_k) - \mathcal{V}(\Theta; \Theta_k). \qquad (A.8)$$

This is due to the fact that

$$\sum_{\mathcal{Z}} \log p(\mathcal{D}; \Theta) p(\mathcal{Z}|\mathcal{D}; \Theta_k) = \log p(\mathcal{D}; \Theta). \qquad (A.9)$$

From Eq. (A.8), one gets

$$\log p(\mathcal{D}; \Theta_{k+1}) - \log p(\mathcal{D}; \Theta_k) = (\mathcal{Q}(\Theta_{k+1}; \Theta_k) - \mathcal{Q}(\Theta_k; \Theta_k))$$
$$- (\mathcal{V}(\Theta_{k+1}; \Theta_k) - \mathcal{V}(\Theta_k; \Theta_k)) \quad (A.10)$$

The second term on the right hand side of Eq. (A.10) can be reformulated as

$$(\mathcal{V}(\Theta; \Theta_k) - \mathcal{V}(\Theta_k; \Theta_k)) = E\left\{\log\left(\frac{p(\mathcal{Z}|\mathcal{D}; \Theta)}{p(\mathcal{Z}|\mathcal{D}; \Theta_k)}\right)|\mathcal{D}; \Theta_k\right\}$$
$$\leq \log E\left\{\frac{p(\mathcal{Z}|\mathcal{D}; \Theta)}{p(\mathcal{Z}|\mathcal{D}; \Theta_k)}|\mathcal{D}; \Theta_k\right\}$$
$$= \log \sum_{\mathcal{Z}} \log p(\mathcal{Z}|\mathcal{D}; \Theta) = 0, \qquad (A.11)$$

where the inequality in Eq. (A.11) is a result of Jensen's inequality [1] and convexity of the logarithmic function. Moreover Eq. (A.11) shows

---

[1] If f is a convex function, then $E\{f(x)\} \geq f(E\{x\})$.

that the second term in left hand side of Eq. (A.10) is non-positive. Hence, selecting a new parameter Θ such that $\mathcal{Q}(\Theta; \Theta_k) \geq \mathcal{Q}(\Theta_k; \Theta_k)$ will lead to $\log p(\mathcal{D}; \Theta) \geq \log p(\mathcal{D}; \Theta_k)$, that means Θ will result in a higher or at least equal likelihood as the Θ_k, which shows that the likelihood value will converge monotonically to some value.

B CSTH Simulink benchmark

The CSTH model developed in [103] consists of MATLAB/SIMULINK based mathematical models combined with experimental data obtained from the real plant. The model utilizes, measured noise and disturbances and provides a realistic benchmark for data-driven analysis and identification.

The SIMULINK model is available from the CSTH benchmark website, http://personal-pages.ps.ic.ac.uk/~nina/CSTHSimulation/index.htm. The CSTH plant is highly nonlinear with its state variables, volume of the water inside the tank and total enthalpy, being nonlinear functions of the input water flow. The thermodynamic properties of

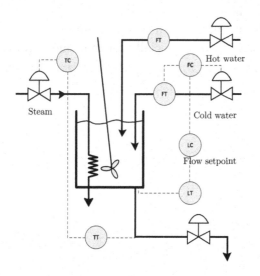

Figure B.1: Continuously stirred tank heater plant

the tank and the output flow also have nonlinear characteristics. The disturbance and noise models are derived from real measurements and the nonlinear behaviors and hard constraints are captured and implemented in look-up tables.

A simple sketch of the plant is shown in Fig. B.1 which consists of a rig in which hot and cold water are mixed and then heated using the stream of steam through the heating coil. The water is then drained from the tank through a long pipe. It is well mixed inside the tank so that the temperature inside the tank can be assumed to be the same as that of the outflow. The plant has three proportional integral controllers for cold water level, temperature and flow.

The plant inputs are hot water, cold water and steam valve position. The level of the cold water and temperature are regulated. The cold water flow, level and temperature of the tank are measured output signals. All the signals are measured in the range of 4-20 mA. The disturbances considered in the model are a deterministic oscillatory disturbance to the cold water flow rate, a random disturbance to the level, and temperature measurement noise, which are obtained from the real pilot plant.

The plant's overall nonlinear behavior and the realistic disturbances offer a challenging task especially for multimode process monitoring approaches.

Bibliography

[1] C.F. Alcala and S.J. Qin. Reconstruction-based contribution for process monitoring. *Automatica*, 45(7):1593–1600, July 2009.

[2] C.F. Alcala and S.J. Qin. Analysis and generalization of fault diagnosis methods for process monitoring. *Journal of Process Control*, 21(3):322–330, March 2011.

[3] E. Alhoniemi, J. Hollmén, O. Simula, and J. Vesanto. Process monitoring and modeling using the self-organizing map. *Integrated Computer-Aided Engineering*, 6(1):3–14, January 1999.

[4] C. Aubrun and D. Sauter. SOA-based platform implementing a structural modelling for large-scale system fault detection: application to a board machine. In *IEEE Multi-Conference on Systems and Control*, Dubrovnik, Croatie, October 2012.

[5] M. Basseville. Detecting changes in signals and systems: a survey. *Automatica*, 24(3):309–326, May 1988.

[6] R.V. Beard. *Failure accomodation in linear systems through self-reorganization*. PhD dissertation, Massachusetts Institute of Technology, 1971.

[7] M. Berrada, S. Tarasiewicz, M.E. Elkadiri, and P.H. Radziszewski. A state model for the drying paper in the paper product industry. *IEEE Transactions on Industrial Electronics*, 44(4):579 –586, August 1997.

[8] Y. Bissessur, E.B. Martin, and A.J. Morris. Monitoring the performance of the paper making process. *Control Engineering Practice*, 7(11):1357–1368, November 1999.

[9] M. Blanke, M. Kinnaert, J Lunze, and M. Staroswiecki. *Diagnosis and Fault-Tolerant Control*. Springer Berlin Heidelberg, 2nd edition, December 2009.

[10] J. Borges, V. Verdult, M. Verhaegen, and M.A. Botto. A switching detection method based on projected subspace classification. In *44th IEEE Conference on Decision and Control*, pages 344 – 349, December 2005.

[11] G. Box, G.M. Jenkins, and G. Reinsel. *Time Series Analysis. Forecasting & Control*. Prentice Hall, 3rd edition, February 1994.

[12] F. Camci. System maintenance scheduling with prognostics information using genetic algorithm. *Reliability, IEEE Transactions on*, 58(3):539 –552, September 2009.

[13] R.V. Canfield. Cost optimization of periodic preventive maintenance. *IEEE Transactions on Reliability*, 35(1):78 –81, April 1986.

[14] J. Chen and R. Patton. *Robust Model-Based Fault Diagnosis for Dynamic Systems*. Kluwer Academic Publishers, 1999.

[15] T. Chen, J. Morris, and E. Martin. Probability density estimation via an infinite gaussian mixture model: Application to statistical process monitoring. *Journal of the Royal Statistical Society. Series C (Applied Statistics)*, 55(5):699–715, January 2006.

[16] T. Chen and Y. Sun. Probabilistic contribution analysis for statistical process monitoring: A missing variable approach. *Control Engineering Practice*, 17(4):469–477, April 2009.

[17] T. Chen and J. Zhang. On-line multivariate statistical monitoring of batch processes using gaussian mixture model. *Computers & Chemical Engineering*, 34(4):500–507, April 2010.

[18] H. Cheng, M. Nikus, and S.L. Jämsä-Jounela. Causal model based fault diagnosis applied on a paper machine simulator. In *1st IFAC Workshop on Applications of Large Scale Industrial Systems*, pages 214–219, Finland, August 2006.

[19] H. Cheng, M. Nikus, and S.L. Jämsä-Jounela. Evaluation of PCA methods with improved fault isolation capabilities on a paper machine simulator. *Chemometrics and Intelligent Laboratory Systems*, 92(2):186–199, July 2008.

[20] L.H. Chiang, E.L. Russell, and R.D. Braatz. Fault diagnosis in chemical processes using fisher discriminant analysis, discriminant partial least squares, and principal component analysis. *Chemometrics and Intelligent Laboratory Systems*, 50(2):243–252, March 2000.

[21] H.W. Cho and K.J. Kim. Fault diagnosis of batch processes using discriminant model. *International Journal of Production Research*, 42(3):597–612, 2004.

[22] S.W. Choi, C. Lee, J.M. Lee, J.H. Park, and I.B. Lee. Fault detection and identification of nonlinear processes based on kernel PCA. *Chemometrics and Intelligent Laboratory Systems*, 75(1):55–67, January 2005.

[23] S.W. Choi and I.B. Lee. Nonlinear dynamic process monitoring based on dynamic kernel PCA. *Chemical Engineering Science*, 59(24):5897–5908, December 2004.

[24] E. Chow and A.S. Willsky. Analytical redundancy and the design of robust failure detection systems. *IEEE Transactions on Automatic Control*, 29(7):603 – 614, July 1984.

[25] B.S. Dayal and J.F. MacGregor. Improved PLS algorithms. *Journal of Chemometrics*, 11(1):73–85, 1997.

[26] R. Dekker. Applications of maintenance optimization models
     a review and analysis. *Reliability Engineering & System Safety*
     51(3):229–240, March 1996.

[27] A.P. Dempster, N.M. Laird, and D.B. Rubin. Maximum likeli-
     hood from incomplete data via the EM algorithm. *Journal of the
     Royal Statistical Society. Series B (Methodological)*, 39(1):1–38
     January 1977.

[28] J. Deng and B. Huang. Identification of nonlinear parameter
     varying systems with missing output data. *AIChE Journal*
     58(11):3454–3467, 2012.

[29] S.X. Ding. *Model-based Fault Diagnosis Techniques: Design
     Schemes, Algorithms, and Tools*. Springer, 1st edition, April
     2008.

[30] S.X. Ding, P. Zhang, E. Ding, S. Yin, A. Naik, P. Deng, and
     W. Gui. On the application of PCA technique to fault diagnosis
     Tsinghua Science & Technology, 15(2):138–144, April 2010.

[31] S.X. Ding, P. Zhang, B. Huang, and E.L. Ding. Subspace
     method aided data-driven design of observer based fault detec-
     tion systems. In *Proceedings of the 16th IFAC World Congress*
     pages 1829–1829, Czech Republic, July 2005.

[32] S.X. Ding, P. Zhang, A. Naik, E.L. Ding, and B. Huang. Sub-
     space method aided data-driven design of fault detection and
     isolation systems. *Journal of Process Control*, 19(9):1496–1510
     October 2009.

[33] G.A. Dumont. Application of advanced control methods in the
     pulp and paper industry: A survey. *Automatica*, 22(2):143–153
     March 1986.

[34] R. Dunia and S.J. Qin. Subspace approach to multidimen
     sional fault identification and reconstruction. *AIChE Journal*
     44(8):1813–1831, 1998.

[35] Eurostat. Pulp, paper and paper product statistics - NACE rev. 1.1. http://epp.eurostat.ec.europa.eu/statistics_explained/index. php/Pulp,_paper_and_paper_product_statistics_-_NACE_Rev. _1.1.

[36] M. Figueiredo and A.K. Jain. Unsupervised learning of finite mixture models. *IEEE Transactions on Pattern Analysis and Machine Intelligence*, 24(3):381–396, March 2002.

[37] P.M. Frank. Fault diagnosis in dynamic systems using analytical and knowledge-based redundancy: A survey and some new results. *Automatica*, 26(3):459–474, May 1990.

[38] P.M. Frank, S.X. Ding, and T. Marcu. Model-based fault diagnosis in technical processes. *Transactions of the Institute of Measurement and Control*, 22(1):57–101, March 2000.

[39] Z. Ge and Z. Song. Multimode process monitoring based on bayesian method. *Journal of Chemometrics*, 23(12):636650, 2009.

[40] Z. Ge, M. Zhang, and Z. Song. Nonlinear process monitoring based on linear subspace and bayesian inference. *Journal of Process Control*, 20(5):676–688, June 2010.

[41] J. Gertler. Residual generation from principal component models for fault diagnosis in linear systems part i: Review of static systems. In *Proceedings of the IEEE International Symposium on Intelligent Control*, pages 634–639, Limassol, Cyprus, June 2005.

[42] J. Gertler. Residual generation from principal component models for fault diagnosis in linear systems part II: extension to optimal residuals and dynamic systems. In *Proceedings of the IEEE International Symposium on Intelligent Control*, pages 634–639, Limassol, Cyprus, June 2005.

[43] J. Gertler, W. Li, Y. Huang, and T. McAvoy. Isolation enhanced principal component analysis. *AIChE Journal*, 45(2):323–334, 1999.

[44] A. Haghani, S.X Ding, J. Esch, and H. Hao. Data-driven quality monitoring and fault detection for multimode nonlinear processes. In *51st IEEE Conference on Decision and Control*, Maui, Hawaii, December 2012.

[45] A. Haghani, S.X Ding, H. Hao, S. Yin, and T. Jeinsch. An approach for multimode dynamic process monitoring using Bayesian inference. In *8th IFAC Symposium on Fault Detection, Supervision and Safety of Technical Processes*, Mexico City, August 2012.

[46] A. Haghani, S.X Ding, T. Jeinsch, H. Hao, and H. Luo. MAP criterion for condition-based maintenance in industrial processes. In *2013 Conference on Control and Fault-Tolerant Systems (SysTol)*, pages 413–418, 2013.

[47] Q.P. He, S.J. Qin, and J. Wang. A new fault diagnosis method using fault directions in fisher discriminant analysis. *AIChE Journal*, 51(2):555–571, 2005.

[48] I.S. Helland. On the structure of partial least squares regression. *Communications in Statistics - Simulation and Computation*, 17(2):581–607, 1988.

[49] H. Holik. *Handbook of Paper and Board*. John Wiley & Sons, October 2006.

[50] A. Höskuldsson. PLS regression methods. *Journal of Chemometrics*, 2(3):211–228, June 1988.

[51] C.C Hsu, M.C Chen, and L.S Chen. A novel process monitoring approach with dynamic independent component analysis. *Control Engineering Practice*, 18(3):242–253, March 2010.

[52] B. Huang. Bayesian methods for control loop monitoring and diagnosis. *Journal of Process Control*, 18(9):829–838, October 2008.

[53] S.A. Imtiaz, S.L. Shah, R. Patwardhan, H. Palizban, and J. Ruppenstein. Development of online monitoring scheme for prediction and diagnosis of sheet-break in a pulp and paper. In *6th IFAC Symposium on Fault Detection, Supervision and Safety of Technical Processes*, pages 837–842, P.R. China, August 2006.

[54] S.A. Imtiaz, S.L. Shah, R. Patwardhan, H.A. Palizban, and J. Ruppenstein. Detection, diagnosis and root cause analysis of sheet-break in a pulp and paper mill with economic impact analysis. *The Canadian Journal of Chemical Engineering*, 85(4):512525, 2007.

[55] R. Isermann. Process fault detection based on modeling and estimation methods: A survey. *Automatica*, 20(4):387–404, July 1984.

[56] R. Isermann and P. Ball. Trends in the application of model-based fault detection and diagnosis of technical processes. *Control Engineering Practice*, 5(5):709–719, May 1997.

[57] S.-L Jämsä-Jounela, V.-M. Tikkala, A. Zakharov, O. Pozo Garcia, H. Laavi, T. Myller, T. Kulomaa, and V. Hämäläinen. Outline of a fault diagnosis system for a large-scale board machine. *The International Journal of Advanced Manufacturing Technology*, June 2012.

[58] S.L. Jämsä-Jounela. Future trends in process automation. *Annual Reviews in Control*, 31(2):211–220, 2007.

[59] L. Jiang, L. Xie, and S. Wang. Fault diagnosis for batch processes by improved multi-model fisher discriminant analysis. *Chinese Journal of Chemical Engineering*, 14(3):343–348, June 2006.

[60] X. Jin and B. Huang. Robust identification of piece-wise/switching autoregressive exogenous process. *AIChE Journal*, 56(7):1829–1844, November 2009.

[61] X. Jin and B. Huang. Identification of switched markov autoregressive eXogenous systems with hidden switching state. *Automatica*, 48(2):436–441, February 2012.

[62] X. Jin, B. Huang, and D.S. Shook. Multiple model LPV approach to nonlinear process identification with EM algorithm *Journal of Process Control*, 21(1):182–193, January 2011.

[63] H.L. Jones. *Failure detection in linear systems*. PhD dissertation Massachusetts Institute of Technology, 1973.

[64] B. Jose, V. Verdult, and M. Verhaegen. Iterative subspace identification of piecewise linear systems. In *14th IFAC Symposium on System Identification*, pages 368–373, Australia, March 2006

[65] A.L. Juloski, S. Weiland, and W. Heemels. A bayesian approach to identification of hybrid systems. *IEEE Transactions on Automatic Control*, 50(10):1520–1533, October 2005.

[66] M. Kano, K. Miyazaki, S. Hasebe, and I. Hashimoto. Inferential control system of distillation compositions using dynamic partial least squares regression. *Journal of Process Control*, 10(2-3):157–166, April 2000.

[67] M. Kano, K. Nagao, S. Hasebe, I. Hashimoto, H. Ohno R. Strauss, and B. Bakshi. Comparison of statistical process monitoring methods: application to the eastman challenge problem. *Computers & Chemical Engineering*, 24(2-7):175–181, July 2000.

[68] M. Kano, S. Tanaka, S. Hasebe, I. Hashimoto, and H. Ohno Monitoring independent components for fault detection. *AIChE Journal*, 49(4):969–976, 2003.

[69] M. Karlsson and S. Stenström. Static and dynamic modeling of cardboard drying part 1: Theoretical model. *Drying Technology*, 23(1-2):143–163, February 2005.

[70] D. Kim and I.B. Lee. Process monitoring based on probabilistic PCA. *Chemometrics and Intelligent Laboratory Systems*, pages 109–123, 2003.

[71] T. Kohonen, E. Oja, O. Simula, A. Visa, and J. Kangas. Engineering applications of the self-organizing map. *Proceedings of the IEEE*, 84(10):1358 –1384, October 1996.

[72] T. Komulainen, M. Sourander, and S.L. Jämsä-Jounela. An online application of dynamic PLS to a dearomatization process. *Computers & chemical engineering*, 28(12):2611–2619, 2004.

[73] W. Ku, R.H. Storer, and C. Georgakis. Disturbance detection and isolation by dynamic principal component analysis. *Chemometrics and Intelligent Laboratory Systems*, 30(1):179–196, November 1995.

[74] W.E. Larimore. Canonical variate analysis in identification, filtering, and adaptive control. In *Proceedings of the 29th IEEE Conference on Decision and Control*, Honolulu, Hawaii, 1990.

[75] J.M Lee, S.J. Qin, and I.B. Lee. Fault detection and diagnosis based on modified independent component analysis. *AIChE Journal*, 52(10):3501–3514, 2006.

[76] J.M Lee, C. Yoo, and I.B Lee. Statistical monitoring of dynamic processes based on dynamic independent component analysis. *Chemical Engineering Science*, 59(14):2995–3006, July 2004.

[77] G. Li, C.F. Alcala, S.J. Qin, and D. Zhou. Generalized reconstruction-based contributions for output-relevant fault diagnosis with application to the Tennessee-Eastman process. *IEEE Transactions on Control Systems Technology*, 19(5):1114–1127, September 2011.

[78] L. Ljung. *System Identification: Theory for the User*. Prentice Hall, 2nd edition, January 1999.

[79] D.G. Luenberger. Observing the state of a linear system. *IEEE Transactions on Military Electronics*, 8(2):74 –80, April 1964.

[80] M.A Massoumnia and W.E. Van der Velder. Generating parity relations for detecting and identifying control system component failures. *Journal of Guidance, Control, and Dynamics*, 11:60–665, February 1988.

[81] G.J. McLachlan and T. Krishnan. *The EM algorithm and extensions*. John Wiley and Sons, February 2008.

[82] G.J. McLachlan and D. Peel. *Finite mixture models*. John Wiley and Sons, September 2000.

[83] P.S. Miller, R. Swanson, and C. Heckler. Contribution plots: A missing link in multivariate quality control. *APPLIED MATHEMATICS AND COMPUTER SCIENCE*, 8(4):775–792, 1998.

[84] A. Naik. *Subspace based data-driven designs of fault detection systems*. PhD thesis, Universität Duisburg-Essen, Duisburg, December 2010.

[85] A. Naik, S. Yin, S.X. Ding, and P. Zhang. Recursive identification algorithms to design fault detection systems. *Journal of Process Control*, 20(8):957–965, September 2010.

[86] H. Nakada, K. Takaba, and T. Katayama. Identification of piecewise affine systems based on statistical clustering technique. *Automatica*, 41(5):905–913, May 2005.

[87] P. Nomikos and J.F. MacGregor. Multivariate SPC charts for monitoring batch processes. *Technometrics*, 37(1):41–59, 1995.

[88] R. Patton and P.M. Frank. *Fault Diagnosis in Dynamic Systems Theory and Application*. Prentice Hall, 1st edition, November 1989.

[89] F. Qi and B. Huang. Bayesian methods for control loop diagnosis in the presence of temporal dependent evidences. *Automatica*, 47(7):1349–1356, July 2011.

[90] F. Qi, B. Huang, and E.C. Tamayo. A bayesian approach for control loop diagnosis with missing data. *AIChE Journal*, 56(1):179–195, January 2010.

[91] S.J. Qin. Partial least squares regression for recursive system identification. In *Proceedings of 32nd Conference on Decision and Control*, San Antonio, Texas, December 1993.

[92] S.J. Qin. Statistical process monitoring: basics and beyond. *Journal of Chemometrics*, 17(8-9):480–502, 2003.

[93] J. Ragot, G. Mourot, and D. Maquin. Parameter estimation of switching piecewise linear system. In *42nd IEEE Conference on Decision and Control*, volume 6, pages 5783 – 5788, December 2003.

[94] A Raich and A. Çinar. Statistical process monitoring and disturbance diagnosis in multivariable continuous processes. *AIChE Journal*, 42(4):995–1009, April 1996.

[95] J. Roll. *Local and piecewise affine approaches to system identification*. PhD dissertation, Linköping university, Department of electrical engineering, Linköping, Sweden, 2003.

[96] E.L. Russell, L.H. Chiang, and R.D. Braatz. *Data-driven Methods for Fault Detection and Diagnosis in Chemical Processes*. Springer, 1st edition, March 2000.

[97] A. Sachdeva, D. Kumar, and P. Kumar. Planning and optimizing the maintenance of paper production systems in a paper plant. *Computers & Industrial Engineering*, 55(4):817–829, November 2008.

[98] M. Sjöström, S. Wold, W. Lindberg, J. Persson, and H. Martens. A multivariate calibration problem in analytical chemistry solved by partial least-squares models in latent variables. *Analytica Chimica Acta*, 150:61–70, 1983.

[99] O. Slätteke. *Modeling and Control of the Paper Machine Drying Section*. Ph.D. dissertation, Department of Automatic Control, Lund University, Lund, Sweden, January 2006.

[100] T. Sorsa, H.N. Koivo, and R. Korhonen. Application of neural network in the detection of breaks in a paper machine. In *Preprints of the IFAC Symp. on On-Line Fault Detection and Supervision in the Chemical Process Industries*, pages 162–167, April 1992.

[101] Stadt Krefeld. Erste Ergebnisse zur Brandursache liegen vor. http://www.krefeld.de/C1257455004E4FBF/html/0BC60C17FE4445BDC1257A86004405B3?Opendocument.

[102] N. Stirken and S. Peters. Zweistelliger millionenschaden. http://www.rp-online.de/niederrhein-sued/krefeld/nachrichten/zweistelliger-millionenschaden-1.3013439.

[103] N.F. Thornhill, S.C. Patwardhan, and S.L. Shah. A continuous stirred tank heater simulation model with applications. *Journal of Process Control*, 18(3-4):347–360, March 2008.

[104] V.-M Tikkala and S.-L. Jämsä-Jounela. Monitoring of caliper sensor fouling in a board machine using self-organising maps. *Expert Systems with Applications*, 39(12):11228–11233, September 2012.

[105] M.E. Tipping and C.M. Bishop. Probabilistic principal component analysis. *Journal of the Royal Statistical Society. Series B (Statistical Methodology)*, 61(3):611–622, January 1999.

[106] P. Van Overschee and B. De Moor. N4SID: Subspace algorithms for the identification of combined deterministic-stochastic systems. *Automatica*, 30(1):75–93, January 1994.

[107] P. Van Overschee and B. De Moor. *Subspace Identification for Linear Systems: Theory - Implementation - Applications.* Kluwer Academic Publishers, 1st edition, May 1996.

[108] V. Venkatasubramanian. A review of process fault detection and diagnosis part I: Quantitative model-based methods. *Computers & Chemical Engineering*, 27(3):293–311, March 2003.

[109] V. Venkatasubramanian, R. Rengaswamy, and S.N. Kavuri. A review of process fault detection and diagnosis part II: Qualitative models and search strategies. *Computers & Chemical Engineering*, 27(3):313–326, March 2003.

[110] V. Venkatasubramanian, R. Rengaswamy, S.N. Kavuri, and K. Yin. A review of process fault detection and diagnosis part III: process history based methods. *Computers & Chemical Engineering*, 27(3):327–346, March 2003.

[111] V. Verdult and M. Verhaegen. Subspace identification of piecewise linear systems. In *43rd IEEE Conference on Decision and Control*, volume 4, pages 3838 – 3843, December 2004.

[112] M. Verhaegen. Subspace model identification part 3: Analysis of the ordinary output-error state-space model identification algorithm. *International Journal of Control*, 58(3):555–586, 1993.

[113] M. Verhaegen and P. Dewilde. Subspace model identification part 1. the output-error state-space model identification class of algorithms. *International Journal of Control*, 56(5):1187–1210, 1992.

[114] M. Verhaegen and P. Dewilde. Subspace model identification part 2. analysis of the elementary output-error state-space

model identification algorithm. *International Journal of Control*, 56(5):1211–1241, 1992.

[115] J. Wang and S.J. Qin. A new subspace identification approach based on principal component analysis. *Journal of Process Control*, 12(8):841–855, December 2002.

[116] J. Wang, A. Sano, T. Chen, and B. Huang. Identification of Hammerstein systems without explicit parameterisation of non-linearity. *International Journal of Control*, 82(5):937–952, 2009.

[117] J.A. Westerhuis, S.P. Gurden, and A.K. Smilde. Generalized contribution plots in multivariate statistical process monitoring. *Chemometrics and Intelligent Laboratory Systems*, 51(1):95–114, May 2000.

[118] D. Westwick and M. Verhaegen. Identifying MIMO Wiener systems using subspace model identification methods. *Signal Processing*, 52(2):235–258, July 1996.

[119] A.S. Willsky. A survey of design methods for failure detection in dynamic systems. *NASA STI/Recon Technical Report N*, 76:601–611, November 1975.

[120] B.M. Wise and N.B. Gallagher. The process chemometrics approach to process monitoring and fault detection. *Journal of Process Control*, 6(6):329–348, December 1996.

[121] S. Wold. Cross-validatory estimation of the number of components in factor and principal components models. *Technometrics*, 20(4):397–405, November 1978.

[122] Y. Xiong and D.Y. Yeung. Mixtures of ARMA models for model-based time series clustering. In *IEEE International Conference on Data Mining*, 2002.

[123] Y. Xiong and D.Y. Yeung. Time series clustering with ARMA mixtures. *Pattern Recognition*, 37(8):1675–1689, August 2004.

[124] R. Xu and D. Wunsch. Survey of clustering algorithms. *IEEE Transactions on Neural Networks*, 16(3):645–678, May 2005.

[125] S. Yin, S.X. Ding, A Haghani, and H Hao. Data-driven monitoring for stochastic systems and its application on batch process. *International Journal of Systems Science*, pages 1–11, 2012.

[126] S. Yin, S.X. Ding, A. Haghani, H. Hao, and P. Zhang. A comparison study of basic data-driven fault diagnosis and process monitoring methods on the benchmark tennessee eastman process. *Journal of Process Control*, 22(9):1567–1581, October 2012.

[127] S. Yin, S.X. Ding, A. Naik, P. Deng, and A. Haghani. On PCA-based fault diagnosis techniques. In *2010 Conference on Control and Fault-Tolerant Systems (SysTol)*. IEEE, October 2010.

[128] S. Yin, S.X. Ding, P. Zhang, A. Haghani, and A. Naik. Study on modifications of PLS approach for process monitoring. In *Proceedings of the 18th IFAC World Congress*, Milano Italy, August 2011.

[129] Q. Yongsheng, W. Pu, and G. Xuejin. Enhanced batch process monitoring and quality prediction using multi-phase dynamic PLS. In *30th Chinese Control Conference (CCC)*, July 2011.

[130] J. Yu. Fault detection using principal components-based gaussian mixture model for semiconductor manufacturing processes. *IEEE Transactions on Semiconductor Manufacturing*, 24(3):432–444, August 2011.

[131] J. Yu. Localized fisher discriminant analysis based complex chemical process monitoring. *AIChE Journal*, 57(7):1817–1828, 2011.

[132] J. Yu. Local and global principal component analysis for process monitoring. *Journal of Process Control*, 22(7):1358–1373, August 2012.

[133] J. Yu. Multiway discrete hidden markov model-based approach for dynamic batch process monitoring and fault classification. *AIChE Journal*, 58(9):27142725, 2012.

[134] J. Yu. Semiconductor manufacturing process monitoring using gaussian mixture model and bayesian method with local and nonlocal information. *IEEE Transactions on Semiconductor Manufacturing*, 25(3):480 –493, August 2012.

[135] J. Yu and S.J. Qin. Multimode process monitoring with Bayesian inference-based finite Gaussian mixture models. *AIChE Journal*, 54(7):1811–1829, July 2008.

[136] J. Yu and S.J. Qin. Multiway Gaussian mixture model based multiphase batch process monitoring. *Industrial & Engineering Chemistry Research*, 48(18):8585–8594, 2009.

[137] P. Zhang and S.X. Ding. Disturbance decoupling in fault detection of linear periodic systems. *Automatica*, 43(8):1410–1417, August 2007.

[138] Y. Zhang and S.J. Qin. Improved nonlinear fault detection technique and statistical analysis. *AIChE Journal*, 54(12):3207–3220, 2008.

[139] D. Zhou, G. Li, and S.J. Qin. Total projection to latent structures for process monitoring. *AIChE Journal*, 56:168–178, 2010.